Python for Beginners

The Ultimate Guide for Data Science and

How to Programming Python Smart Way to

Learn Data Analytics & Deep Learning Faster

Computer Programming for Beginners

Tony F. Charles

Table of Contents

Introduction

Python is a great language to start your programming journey. It has an easy syntax that makes sense to non-programming professionals. In the background, it's based on the C language which makes it extremely powerful. Since created by Van Rossum in 1991, Python has remained an open source project. It is for this reason you will find a lot of online tutorials and resources about Python.

Today, more fields than ever before are replying on computers to achieve results. Whether you are an aspiring social worker or a game designer, you require a particular degree of computer knowledge to get employment. Even if it's not a requirement, you can increase your productivity by automating your daily tasks. Consider you are a social worker who has to send regular email updates to your boss about the cases you work everyday. Since you already record notes on your computer, you might be able to write a script to extract information from your notes and send emails to your boss automatically at the end of your day. This is just one example, and I am sure you can think about a dozen similar tasks that you do that can be automated.

If you are starting with programming and have no knowledge about computers and programming, this book is written for you. You will find initial chapters easier to understand and as you progress to later chapters, you will have better skills and we will cover more advanced topics. This book is written in one flow and it's better to not skip any topic, let alone an entire chapter.

The software industry is changing rapidly and by the time you are finished with this book, there might be better techniques or tools available. This book focuses on using multiple tools to get the same results so the following can be instilled in your mind: "It doesn't matter what tools you use, the only thing that makes headlines is the end-product." I have high hopes this book will make your programming journey easier.

Chapter 1: Computer Science and Mathematics

Computers were developed by scientists to solve problems. Mathematics is the scientific language to solve problems. Therefore, computers and mathematics are inseparable. Computers utilize mathematical concepts to identify and solve problems. They are much better in sequential processing and more recently in multitasking different processes. Take the example of a french fries seller. He peels and cuts 50 potatoes all by himself. With time, more people start to buy fries from him and he needs to peel and cut more potatoes. He can hire helpers who he will have to pay per hour and also afford their medical and accommodate for their availability. The other option is to get a computer aided automated machine that takes the potatoes, washes, peels and then cuts them according to given criteria. The initial setup cost might be high but due to low operational cost, this is the better option. Not only the cost, the machine will be able to process more potatoes much quicker compared to human helpers.

This is how computers are changing everyone's lives. The cellphone you use all the time is also a computer, a small yet very sophisticated and powerful computer. Let's look at the history of computers.

History of Computers

In the 1880 US census, the population was determined to be a little over 50 million, 50,189,209 to be exact. It took 7 years for the US government to compile the census results and come up with the total number. There was definitely a need to process the census data faster. This need resulted in the invention of punch card computers that were gigantic machines. But, the concept of computers even predates this era.

Abacus

Not a computer but a counting tool, Abacus has been used for centuries by many ancient civilizations to count. How the tool came into existence is a complete mystery. It is still used today to teach children and visually impaired persons how to count.

Abacus is based upon the concept of coding - representing values with symbols. In the case of abacus, numbers are represented by beads and their values are determined by their positions.

1801: To automate the weaving process of different fabric designs, Joseph Marie Jacquard invented a punch card reading loom. This punch card reading technology was later used by electronic computers in the early 20th century.

1822: People have hated memorizing multiplication tables for ages. Want proof? Charles Babbage, an English mathematician, invented a steam-powered machine to calculate the number tables. The English government funded project ultimately failed, but laid the foundation for computers developed later on.

1890: Remember when we said it took US government seven years to process the 1880 census? In the 1890 census, Herman Hollerith solved the issue by designing a system based on punch cards. It saved $5 million for the government and completed the process in only three years. The company Herman founded later became IBM.

1936: The modern computer is a good example of a Turing machine, a concept coined by Alan Turing in 1936 about a machine that can compute anything that's possible to compute.

1937: Up till this year, computers were electro-mechanical machines. A physics and mathematics professor at Iowa State,

J.V. Atanasoff, attempted to build a computer without moving mechanical parts. Did he succeed? Time to do some Googling!

1939: The now globally recognized tech-giant, manufacturers of computer systems and peripherals, the "hp" company was founded by Bill Hewlett and David Packard in a garage of the city of Palo Alto, California.

1941: With the help of his graduate student Clifford Berry, J.V. Atanasoff one ups his previous achievement by designing a computer that could simultaneously solve 29 equations. The computer was the first of its kind because it could hold information in its memory.

1943-44: Two University of Pennsylvania professors, J.P. Eckert and John Mauchly, invent the Electronic Numerical Integrator and Calculator (ENIAC). A precursor to digital computers, it was physically enormous comprising of more than 17,500 vacuum tubes in a 20ft by 40ft assembly.

1946: The United States Census Bureau funded a contract for Mauchly and Eckert to build Universal Automatic Computer (UNIVAC). UNIVAC became the first commercially available computer and was widely used by government departments and research, financial and aviation institutions.

1947: The era of electronic computers started with the invention of the solid state transistor by William Shockley, Walter Brattain and John Bardeen of Bell laboratories. The transistor worked as a virtual switch that could either be off or on to store information. It was much smaller in size, consumed much less power and generated much less heat compared to vacuum tubes.

1953: COmmon Business-Oriented Language (COBOL), the first programming language, was invented by Grace Hooper. Thomas Watson Jr. invents the IBM 701 EDPM for the sole purpose of enabling the United Nations to monitor the Korean region during the Korean War.

1954: With computers becoming more commonly used by various companies, the need for a better programming language became apparent. A team of IBM programmers headed by John Backus developed a programming language named FORMula TRANslation (FORTRAN).

1958: Thanks to rapid advancements in the field of electronics, Jack Kilby and Robert Noyce developed the first integrated circuit to be used as a computer chip. The Nobel Prize in the field of Physics was awarded to Kilby in 2000, more than 40 years after his incredible feat.

1964: A computer with a mouse and Graphical User Interface (GUI) is demonstrated by Douglas Engelbert as a modern computer prototype. The purpose was to introduce a highly accessible computer model that the general public could feel comfortable with. Computers were no longer just a scientist's play toy.

1962: There was no compatibility and standardization as different manufacturers independently designed computer systems. The issue was addressed by a team of Bell Labs programmers who created UNIX, a C language based operating system. The operating system could be run on various computer systems of that time. The era of system independent software was kickstarted even though UNIX was mostly deployed on mainframe systems by corporations instead of personal computers due to slow performance.

1970: Intel joined the computer race by launching the first temporary memory chips known as Intel 1103 Dynamic Random Access Memory (DRAM).

1973: Connecting multiple computers and/or hardware to combine computational power became easy when one of the research staff members at Xerox, Robert Metcalfe, invented the Ethernet protocol.

1974-77: Personal computers became more popular which led to the unveiling of several different models in the market including IBM 5100, Scelbi & Mark-8 Altair and the Radio Shack's "Trash 80", model TRS-80. Another popular personal computer from that era is the Commodore PET.

1975: The Altair 8080 got featured in the Popular Electronics (PE) magazine as the first true minicomputer for personal use that provide the same computational power of commercial computers. It inspires two computer enthusiasts Paul Allen and

Bill Gates to write software for the popular minicomputer. Both later start a software firm on the back of this successful endeavour, called Microsoft.

1976: Apple Computers went operational led by the Two Steves, Steve Jobs and Steve Wozniak. Their first computer Apple 1 was launched on April Fool's day and was the first computer that used a single circuit board.

1977: The Trash 80 demand rose so high among the general public that Radio Shack had to manufacture more than the initial 3,000 unit production. The multipurpose nature of the computer made it hugely popular. In the same year, the Two Steves incorporated Apple and launched the Apple II during the West Coast Computer Fair. The computer offered color graphics and better storage via a drive for audio cassettes.

1978: The first spreadsheet software, called VisiCalc, got introduced. No one knows that name today!

1979: The first word processing software WordStar was launched by MicroPro International headed by Rob Barnaby. The software had various editing and formatting options including the ability to add margins, word wrap and print documents. The company went with a modernistic approach with the software and did not include a command mode which was usually a feature in software of that era. Another great software of that era that did not make it through the years.

1981: The first personal computer made of IBM's Intel hardware and Microsoft's MS DOS operating system, named "Acorn" was launched. Even today, the majority of personal computers have the combination of IBM and Microsoft products. Acorn offered two floppy disk drives and supported a color monitor. For the first time, the computer was sold through third party distributors and was popularly referred to as the PC.

1983: Apple introduced the first personal computer with a true GUI, named Lisa. The computer interface featured graphical icons and dropdown menus. The product flopped but paved the way for the future Apple computer system line of Macintosh. The same year, Gavilan Computer Corp. unveiled the Gavilan SC, the first widely recognized portable personal computer (laptop).

1985: To counter Apple's Lida, Microsoft introduced Windows which became the most widely used operating system in personal computers. In the same year, Commodore launched the Amiga 1000 that had enhanced audio and video abilities. In the same year, a Massachusetts based computer manufacturer Symbolics, purchased the first dot com website domain in the world: "symbolics.com". At this time, there wasn't even a concept of Internet history.

1986: Compaq launched the Deskpro 386 that had a 32bit architecture offering better performance than the competition to the extent it rivalled the mainframes of that era.

1990: In any organization of this era there were several hundred computers, and there was also a protocol to connect them (ethernet), but there was no good way to present information. A European Organization for Nuclear Research (commonly known as CERN, which is short for the french "Conseil européen pour la recherche nucléaire") researcher, Tim Berners-Lee, developed the HyperText Markup Language (HTML) which later became the language of the World Wide Web.

1993: Intel introduced the revolutionary Pentium processors that enabled high quality graphics and audio services to be offered in personal computers.

1994: The launch of hit video games including "Descent", "Alone in the Dark 2", "Magic Carpet", "Command & Conquer", "Little Big Adventure" and "Theme Park" meant the computers were not just for work, people were actively using them for entertainment. There was a demand for better performing computers.

1996: The search engine most people use today, Google, went live developed by Larry Page and Sergey Brin.

1997: Apple was struggling and in a legal dispute with Microsoft with the allegation that Microsoft copied the design and feel of its operating system. Apple took back the case after Microsoft invested an amount of $150 million in Apple.

1999: More mobility freedom became possible for the personal computer users after WiFi was introduced which meant internet and local connections could be made without the need for physical wires.

2001: The rivalry between Apple and Microsoft intensified after Apple released the newly designed Mac OS X. It was an instant hit as it offered secured memory architecture, as well as superior and faster multitasking by preserving software states. As a response, Microsoft released the Windows XP with a distinct GUI and advanced features.

2003: Albeit having a much smaller market share, AMD one ups Intel this time by launching the first ever 64bit system known as Athlon 64 which became an instant hit.

2004: Mozilla launches Firefox browser to rival the widely used Internet Explorer by Microsoft. Although not very intuitive, Internet Explorer was widely used because it was bundled with the popular Windows operating system. Mark Zuckerberg launched Facebook which was a social media platform.

2005: YouTube, the biggest video sharing platform today, was designed and launched by three PayPal employees, Jawed Karim, Steve Chen and Chad Hurley. Google acquired the Linux-based operating system for mobile devices known as Android.

2006: Apple designed the first mobile computer (laptop) MacBook Pro and the desktop computer iMac both of which used an Intel-based dual core system. Nintendo Wii gaming console became available in the market that had games based on motion and gesture detection.

2007: Apple released the groundbreaking smartphone with the brand name iPhone which remained the best-selling cell phone for a long time.

2009: Windows 7 was launched by Microsoft which became widely popular and had advanced capabilities including touch and speech recognition.

2010: Apple reignited consumer interest in tablets by launching the iPad.

2011: Google released the Chrome OS based laptop with the brand name Chromebook.

2012: Facebook gained 1 billion users, becoming the most significant social media platform.

2015: Apple released another technically advanced device: the Apple Watch. Microsoft released the final version of Windows, Windows 10, but still kept support for Windows 7 for years due to its wide use.

2016: Not the first quantum computer, but the first reprogrammable quantum computer was created. Before this, quantum computers were manufactured for specific purposes.

2017: Big improvements were made in the field of "molecular informatics", the science of creating computers using molecules as processing units.

The advancements in computer systems grow everyday. The silicon based electronics have almost reached their limits. There are various researches going on to find a better base for electronic devices, but they are still far from commercial use.

Computer Architecture

The science of joining computer hardware and software to create unique computational systems is called computer architecture. It has evolved many times during the years as the applications of computers changed. Von Neumann architecture is the most famous computer architecture that relates to modern digital computers. According to this architecture, all computer systems are more or less comprised of following main parts.

Central Processing Unit (CPU)

The brain of a computer, the central processing unit, processes the given input and generates the desired output. The CPU itself is comprised of various parts.

Control Unit

Control Unit controls all the aspects of the CPU so the latter can process data according to set instructions. The supported set of instructions are stored in the permanent memory (Read Only Memory - ROM). It also controls the flow of data within the computer and provides signals for timing operations. If you are a gamer or a professional who has to deal with high quality graphics, you might have tweaked the clock speed of your computer's graphics unit. This is direct interaction with the computer's control unit.

Processor

The processor performs the actual computations. It houses the logic circuits and onboard caches, known as registers, for fast temporary data storage. Registers, which should not be confused with the main memory (RAM), enable the processor to store intermediate data generated during complex computations. The logic circuitry has electronic gates that operate on given data in different ways as told by the given instructions.

Main Memory

The Random Access Memory (RAM) is referred to as the main memory of a computer. Why? Because all the currently running applications are kept here until something is no longer needed. Modern computers have very large RAMs which enables them to hold many dormant applications for a longer time which makes multitasking and application boot up time very fast. Failure of power usually erases the contents of the RAM.

Input Systems

There are various input peripherals that have been developed for different purposes. The most common is the keyboard and the mouse. We have now touch screens, touch pads, voice (microphones) and image recognition (cameras) built into the computer. You can also use a scanner to scan hardcopy documents for easy processing and transmission.

Output Systems

Just like input devices, there are several output peripherals, the most common being the monitor and speaker. The other commonly known output device is the printer. It's interesting to note that the output devices are very few as compared to the input devices. It shows that the focus of computer architecture has always been to enable computers to take input so they can process it in the desired way.

More Components in A Modern Computer

A modern computer has a lot of different components added for enhanced processing capabilities.

External Memory

If you look at the computer architecture, you would notice there's no storage available for the user to store any information. The ROM is read only and houses the instructions and the RAM is temporary. In earlier computers, there were floppy disk drives that acted as an external memory so the computer user could save any information. In later computers, a hard disk was introduced as a permanent part of the computer system that was used to store all the information. With the advancement in memory storage, more devices were introduced such as CDs, DVDs and now the most common, USB flash drives.

Graphics Processing Unit (GPU)

Modern computers are frequently used in applications that demand high quality graphics. Video games have become very common, professionals have to deal with creation and editing of high quality graphics. Even watching high definition movies takes a lot of processing power. To remove excessive load from the CPU of a computer, a standalone processing unit is added to handle all graphics related tasks. Not only does it divide the load, it also divides the heat generation so one part of the computer doesn't get too hot.

<u>Virtual Memory (Paging)</u>

Applications these days take a lot of memory. What happens if your computer runs out of available RAM? Does it hang the system? Not so! Every computer has active memory management that monitors the amount of storage required for uninterrupted operation of the computer. If the RAM is full, the computer declares a portion of the hard disk as a virtual memory section and uses it as extended RAM (or virtual cache). The process is called pagination. Of course, this section is slower than the actual RAM, but helps keep applications ready for access. The user can control how much hard disk space can be used for virtual memory.

Mathematical Concepts

As we have stated before, computers are physical manifestations of several mathematical concepts. Mathematics are the scientific language of solving problems. Over the centuries, mathematicians have theoretically solved many complex issues. Mathematics includes concepts like algebra and geometry.

Number Systems

Mathematics is a game of number manipulation which makes number systems at the center stage of mathematical concepts. There are several different types of number systems. Before we take a look at the number systems, we have to understand the concept of coding.

Coding

A way to represent values using symbols is called coding. Coding is as old as humans. Before the number systems we use today, there were other systems to represent values and messages. An example of coding from ancient times is the Egyptian hieroglyphs.

Number systems are also examples of coding because values are represented using special symbols.

There are different types of number systems, and we are going to discuss a few relevant ones.

Binary System

A binary system has only two symbols, 1 and 0 which are referred to as bits. All the numbers are represented by combining these two symbols. Binary systems are ideal for electronic devices because they also have only two states, on or off. In fact, all electronic devices are based on the binary number system. The number system is positional which means the position of symbols determines the final value. Since there are two symbols in this system, the system has a base of 2.

The sole purpose of input and output systems is to convert data to and from binary system to a form that makes better sense to the user. The first bit from the left side is called Most Significant Bit (MSB) while the first bit from the right is called the Least Significant Bit (LSB).

Here is the binary equivalent code of "this is a message":

01110100 01101000 01101001 01110011 00100000 01101001
01110011 00100000 01100001 00100000 01101101 01100101
01110011 01110011 01100001 01100111 01100101

Decimal System

The decimal system has ten symbols, the numbers 0 through 9. This is also a positional number system where the position of symbols changes the value it represents. All the numbers in this system are created with different combinations of the initial ten symbols. This system has a base 10.

This is also called the Hindu-Arabic number system. Decimals make more sense to humans and are used in daily life. There are two reasons for that.

1. Creating large numbers from the base symbols follows a consistent pattern
2. Performing arithmetic operations in a decimal system is easier compared to other systems

Hexadecimal System

The hexadecimal number system is the only one that has letters as symbols. It has the 10 symbols of the decimal system plus the six alphabets A, B, C, D, E and F. This is also a positional number system with a base 16.

Hexadecimal system is extensively used to code instructions in assembly language.

Number System Conversion

We can convert the numbers from one system to another. There are various online tools to do that. Python also offers number conversion, but it is better to learn how it is done manually.

Binary to Decimal

Here's a binary number 01101001, let's convert it to a decimal number.

$(01101001)_2 = 0 \times 2^7 + 1 \times 2^6 + 1 \times 2^5 + 0 \times 2^4 + 1 \times 2^3 + 0 \times 2^2 + 0 \times 2^1 + 1 \times 2^0$

$(01101001)_2 = 0 + 64 + 32 + 0 + 8 + 0 + 0 + 1$

$(01101001)_2 = (105)_{10}$

Decimal to Binary

To convert a decimal number to binary, we have to repeatedly divide the number by two until the quotient becomes one. Recording the remainder generated at each division step gives us the binary equivalent of the decimal number.

2	105	
2	52	1
2	26	0
2	13	0
2	6	1
2	3	0
	1	1

$(105)_{10} = (1101001)_2$

An interesting thing to note here is that $(01101001)_2$ and $(1101001)_2$ represent the same decimal number $(105)_{10}$. It means that just like decimal number system, leading zeros can be ignored in the binary number system.

Binary to Hexadecimal

Binary numbers can be converted to hexadecimal equivalents using two methods.

1. Convert the binary number to decimal, then decimal to hexadecimal number
2. Break binary number in groups of four bits and convert each to its hexadecimal equivalent, keeping the groups' positions in the original binary number intact.

Let's convert $(1101001)_2$ to a hexadecimal number using the second method. The first step is to break the binary number into different groups each of four bits. If the MSB group has less than four bits, make it four by adding leading zeros. Grouping starts from the LSB. So, $(1101001)_2$ will give us $(1001)_2$ and $(0110)_2$. Now, remembering their position in the original binary number, we are going to convert each group to a hexadecimal equivalent.

Here is the table of hexadecimal equivalents of four-bit binary numbers.

Binary	Hexadecimal
0000	0
0001	1
0010	2

0011	3
0100	4
0101	5
0110	6
0111	7
1000	8
1001	9
1010	A
1011	B
1100	C
1101	D
1110	E
1111	F

From the table, we can see $(1001)_2$ is $(9)_{16}$ and $(0110)_2$, the MSB group, is $(6)_{16}$.

Therefore, $(1101001)_2 = (01101001)_2 = (69)_{16}$

Hexadecimal to binary

We can use the above given table to quickly convert hexadecimal numbers to binary equivalents. Let's convert (4EA9)$_{16}$ to binary.

(4)$_{16}$ = (0100)$_2$

(E)$_{16}$ = (1110)$_2$

(A)$_{16}$ = (1010)$_2$

(9)$_{16}$ = (1001)$_2$

So, (4EA9)$_{16}$ = (0100111010101001)$_2$ = (100111010101001)$_2$

Decimal to Hexadecimal

You can say hexadecimal is an extended version of decimals. Let's convert (45781)$_{10}$ to decimal. But, first, we have to remember this table.

Decimal	Hexadecimal
0	0
1	1
2	2
3	3
4	4
5	5

6	6
7	7
8	8
9	9
10	A
11	B
12	C
13	D
14	E
15	F

We are going to divide the decimal number repeatedly by 16 and record the remainders. The final hexadecimal equivalent is formed by replacing remainder decimals with their correct hexadecimal symbols.

16	45781	
16	2861	5
16	178	13
	11	2

$(45781)_{10} = (B2D5)_2$

Hexadecimal to Decimal

Let's convert (4EA9)$_{16}$ to its decimal equivalent.

(4EA9)$_{16}$ = 4 x 16^3 + 14 x 16^2 + 10 x 16^1 + 9 x 16^0

(4EA9)$_{16}$ = 16384 + 3584 + 160 + 9

(4EA9)$_{16}$ = (20137)$_{10}$

There's another number system, the octal system, where the number of unique symbols include 0, 1, 2, 3, 4, 5, 6, along with 7. These were developed for small scale devices that worked on small values with limited resources. With the rapid advancements in storage and other computer resources, octal system became insufficient and thus was discarded in favor of hexadecimal number system. You might still find an old octal based computer system.

Fractions (Floating Points)

Decimal number system supports a decimal point '.' to represent portion/slices of a value. For example, if we want to say half of milk bag is empty using numbers, we can write 0.5 or ½ of milk bag is empty. Do other number systems support decimal point? Yes, they do. Let's see how to convert (0.75)$_{10}$ or (¾)$_{10}$ to binary.

¾ x 2 = 6/4 = 1 . (2/4)

2/4 x 2 = 4/4 = 1

$$(0.75)_{10} = (\tfrac{3}{4})_{10} = (.11)_2$$

Negatives

In the decimal system, a dash or hyphen '-' is placed before a number to declare it as a negative. There are different ways to denote negative numbers in the binary system. The easiest is to consider the MSB as a sign bit, which means if MSB is 1, the number is negative and if the MSB is 0, the number is positive. Determining if a hexadecimal number is negative or positive is a bit tricky. The easiest way is to convert the number into binary and perform the checks for negatives in binary system.

Linear Algebra

Did you hate algebra in school? I have some bad news for you! Linear algebra is heavily involved in programming because it's one of the best mathematical ways to solve problems. According to Wikipedia, algebra is the study of mathematical symbols and the rules for manipulating these symbols. The field advanced thanks to the works of Muhammad ibn Musa al-Khwarizmi who introduced the reduction and balancing methods and treated algebra as an independent field of mathematics. Algebra comes from (*al-jabr*) in the name of the book *al-Kitāb al-mukhtaṣar fī ḥisāb al-jabr wal-muqābala* (The Compendious Book on Calculation by Completion (*al-jabr*) and Balancing (*al-muqābala*)) Khwarizmi wrote on the subject. During that era, the concept of 'x' and 'y' etc. variable notation wasn't widespread but during the Islamic Golden Age, Arabs had a fondness of lengthy "layman terms" descriptions of problems and solutions and that is what Khwarizmi explained algebra concepts in his book. The book dealt with many practical real-life problems including the fields of finance, planning, and legal.

So, we know what algebra is. But, where does "linear" comes from? For that, we have to understand what a linear system is. It is a mathematical model where the system attributes (variables) have a linear relation among themselves. The easiest way to explain this is if the plot between system attributes is a straight line, the system is linear. Linear systems are much simpler than the nonlinear systems. The set of algebraic concepts that relate to linear systems is referred to as linear algebra. Linear algebra helps resolve system problems such as missing attribute values. The first step is to create linear equations to establish the relationship between the system variables.

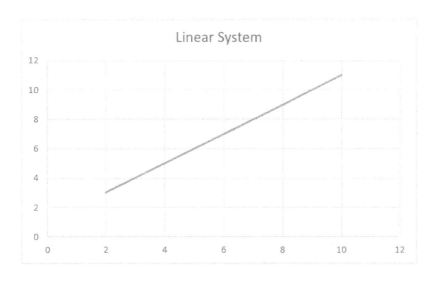

Statistics

Another important field of mathematics that is crucial in various computer science applications. Data analysis and machine learning wouldn't be what they are without the advancements made in statistical concepts during the 20th century. Let's see some concepts related to statistics.

Outlier

Outlier detection is very important in statistical analysis. It helps in homogenizing the sample data. After detecting the outliers, what to do with them is crucial because they directly affect the analysis results. There are many possibilities including:

Discarding Outlier

Sometimes it's better to discard outliers because they have been recorded due to some error. This usually happens where the behaviour of the system is already known.

System Malfunction

But, outliers can also indicate a system malfunction. It is always better to investigate the outliers instead of discarding them straightaway.

Average

Finding the center of a data sample is crucial in statistical analysis because it reveals a lot of system characteristics. There are different types of averages, each signifying something important.

Mean

Mean is the most common average. All the data values are added and divided by the number of data values added together. For example, you sell shopping bags to a well-renowned grocery store and they want to know how much each shopping bag can carry. You completely fill 5 shopping bags with random grocery items and weigh them. Here are the readings in pounds.

5.5, 6.0, 4.95, 7.1, 5.0

You calculate the mean as $(5.5 + 6 + 4.95 + 7.1 + 5) / 5 = 5.71$. You can tell the grocery store your grocery bags hold 5.71 lbs on average.

Median

Median is the center value with respect to the position of data in a sample data when it's sorted in ascending order. If sample data has odd members, the median is the value with an equal number of values on both flanks. If sample data has an even number of values, the median is calculated by finding the mean of two values in the middle with equal number of items on both sides.

Mode

Mode is the most recurring value in a dataset. If there is no recurring value in the sample data, there is no mode.

Variance

To find how much each data value in a sample data changes with respect to the average of the sample data, we calculate the variance. Here is a general formula to calculate variance.

sum of (each data point - mean of sample points)^2 / number of data points in the sample.

If the variance is low in a sample data, it means there are no outliers in the data.

Standard Deviation

We take the square root of variance to find standard deviation. This relates the mean of sample data to the whole of sample data.

Probability

No one can accurately tell what will happen in the future. We can only predict what is going to happen with some degree of certainty. The probability of an event is written mathematically as,

Probability = number of possible ways an event can happen / total number of possibilities

A few points:

1. Probability can never be negative
2. Probability ranges between one and zero

3. To calculate probability, we assume that the set of events we are working with occur independently without any interference.

Finding the probability of an event can change the probability of something happening in a subsequent event. It depends upon how we are interacting with the system to find event probabilities.

Distribution

There are many types of distributions. In this book, whenever we talk about distribution, we refer to probability distribution unless explicitly stated otherwise. Let's take an example of flipping coins and see what is the distribution of such events.

H H H

H H T

H T H

T T T

T H H

H T T

T H T

T T H

This is a very simple event with only a handful of possible outcomes. We can easily determine the probability of different outcomes. But, this is almost impossible in complex systems with thousands or millions of possible outcomes. Distributions work much better in such cases by visually representing the probability curve. It makes more sense than looking at a huge table of fractions or small decimal numbers.

We call a probability distribution discrete if we know all the possible outcomes beforehand.

Chapter 2: Introduction to Python

Before doing anything else, visit https://www.python.org/. The website contains all Python standard packages and comprehensive documentation. Here's a problem: official documentations are written for professional programmers and it doesn't make sense for beginners. This is where Python beats the competition: the huge community of volunteers and contributors who have created so much learning material that's ideal for beginners. This book is also part of that learning material.

Why Python?

Almost everyone who starts learning programming questions the choices they made before, like why did I pick this language? Why I am learning programming? If you went with Python because it's the trend, we can assure you that you have made a correct decision. Python is a high-level language built for beginners. From the start, the language instils the concept of arranged, clean code in the minds of a programmer. In other languages, you spend considerable time learning the syntax. But with Python, you learn how to program without worrying too much about the specific syntax. It doesn't mean there is no specific syntax to follow, it means there are less things to worry about.

The programming language is open-source which means more people have access to it. You will find tutorials on almost everything there is to do in Python. I still remember the last few months of my college days when I was stuck with my final project. I had tried everything I could think of for days but there was no progress. I started searching on the Internet and after some more worryful hours, I stumbled upon the work of another student who had solved a similar issue. I modified his approach to my requirements and voila, my project was complete! I am very sure that if I had been working with a proprietary platform, I would have failed that course.

Remembering this, I should tell you an important skill that you should master before you start learning programming. The skill is to become an advanced user of search engines. No matter which search engine you use, Google, Bing, Yahoo, etc., you must know all or most of the tips and tricks on how to find something online. There is a high chance someone has already gone through what you are trying to do, or has done something similar. You will feel like a hacker once you master the art of using search engines!

Basic Stuff

Constants

Constants are symbols that already have fixed value in the programming language. As the name suggests, you cannot change constants. Constants have specific data type. For example, number 21 is a constant.

Expressions

Expressions will most frequently appear on the right hand side of an operator and can be a combination of constants, variables, and more operators. When writing a script, an expression will not result in any output unless the result of expression is either stored in a variable or used to make a decision.

Operators

A wide range of operators are supported by Python. Operators are commands/instructions that perform specific operation on the data supplied with the operators which are themselves called operands. Here's a list of some important operators in Python.

Arithmetic Operators

Arithmetics is another vital part of mathematics and Python supports a long list of arithmetic operations. Here are some examples.

Operator	Meaning
+	Perform addition of operands - order of operands doesn't matter
-	Perform subtraction of one operand from the other (right side from the left side)

*	Multiply operands and their order doesn't affect the output
/	Divide one operand from the other (left side by the right side) and returns the quotient
%	Divides one operand from the other (left side by the right side) but returns the remainder
**	Returns exponents of an operand

Comparative Operators

These operators are vital for conditional statements. The below table contains some examples.

Operator	Meaning
==	Check if two operands are equal
!=	Check if two operands are not equal
>	Check if the left side operand is greater than the right side operand
<	Check if the left side operand is less than the right side operand

Logical Operators

We can perform multiple operations together and see the combined effect.

Operator	Meaning
and	performs logical AND operation - only true if all operands are true
or	performs logical OR operation - only false if all operands are false
not	reverse the result of a logical operation

Assignment Operator

The only operator that lets you store information, the '=' operator, is very crucial to Python programming. On the left side is always the variable name, and on the right side is a constant, expression, or a function that returns a value.

Python executes operators according to the PEMDAS rule of precedence. Here is the summary.

1. Anything wrapped with parentheses will be executed first. If you want to execute an operation first by ignoring its precedence, put them inside parentheses.
2. Exponent operator has the next highest precedence.

3. The next precedence is shared by multiplication and division operations
4. Last but not least, addition and subtraction have the same precedence and are executed at the very end

If more than one operators have the same precedence, Python will execute the operator that's at the leftmost side of the expression. Here are a few practical examples.

```
>>> 4 + ( 8 * ( 3 - 1 ) ) + ( 12 * 2 )

44

>>> 9183 % 21

6

>>> mon = 767

>>> mon * ( mon / mon )

767.0
```

Variables

You can store constants or results gathered through a function return or expression result in a variable. In Python, we do not have to pre-declare the type of variable. The type of data stored in the variable determines its type.

```
a = 4

print a
```

We stored the integer '4' in the variable 'a'. It makes the variable 'a' of the integer type.

There are different rules you need to follow when creating a variable name.

1. The first character of a variable name must be an alphabet
2. Numbers can be used in a variable name
3. You can capitalize the first name of a variable name but it's not considered good practice
4. Both uppercase and lowercase alphabets can be used in a variable name
5. Underscore '_' is the only special character allowed in a variable name

Python has 31 keywords in total that you cannot use as a variable name. Here's the complete list for reference.

global

2. if
3. import
4. in
5. is
6. lambda
7. not
8. or
9. pass
10. print
11. raise
12. return
13. try
14. while
15. with
16. yield
17. and
18. as
19. assert
20. break
21. class
22. continue
23. def
24. del
25. elif
26. else
27. except
28. exec
29. finally
30. for
31. from

31.

A few examples of assigning variables are below

sTr = "Howdy partner!" #string variable

fLt = 3.14 # floating variable

sTRng = "3.14" #double quotes mean its a string

To check the type of variable, we can use the type() method.

type(sTRng)

You will see the following output.

<type 'str'>

Statements

All the instructions that you give to Python are called statements. Statements may or may not produce an output. When you write a script, you are sequencing a number of statements for the Python interpreter to execute one by one.

Test the following lines of code using Python.

i = 411

i = i + 27

print i

We are doing something unique here. We know that in Python, the right side of the operator '=', which is "i + 27" is calculated first and the result is assigned to the variable name 'i' on the left. Using this knowledge, we have overwritten the value of variable 'i' with a new value.

End of statements

Most programming languages were inspired from the C language, hence use the semicolon ';' to mark the end of a statement. Python, although based on the C language, deviated from this practice and used newlines and tabs to demarcate code blocks. It increased readability but creates problem when copy pasting codes.

If you want to write a single line of code to take two lines on your screen, you can use a backward slash '\'. Here's an example.

```
>>> breKK = \
... 971 + \
... 412
```

The three dots indicate the Python interpreter is waiting for the next part of the code. You will see this when writing a loop, conditional or function code block. Checking the value of variable "breKK" will tell if the program ran correctly. The returned value should be 1383.

```
>>> breKK
```

1383

Conditionals

We use conditionals to check a condition and direct the flow of the program execution. These statements are integral to deploying the concept of control flow in Python.

The if statement

If the given condition is true, execute the next block of code otherwise ignore it.

```
>>> num = 77

>>> if num == 18:

...      print("Execute this")

...

>>>
```

There's no output because the print statement never got executed because the condition "if num == 18" returned false. The indented blocks are referred to as suites. In case the if condition is true, the print statement will be executed. The following code gives the output of "Execute this".

```
>>> num = 77

>>> if num == 77:

...        print("Execute this")

...

Execute this

>>>
```

We can combine the "if" statement with the "else" statement to execute codes in case the "if" condition returns false.

```
>>> vaRa = 47

>>> if vaRa < 23:

...        print("vaRa is less")

... else:

...        print("vaRa is more")

...

vaRa is more

>>>
```

The variable "vaRA" is assigned the value of 47. The "if" condition returns false which directs the control flow to the code block under "else" statement for execution. We can use elif to check a condition before executing a code block.

```
>>> vaRa = 47

>>> if vaRa < 23:

...     print("vaRa is less")

... elif xcon > 33:

...     print("vaRa is more")

... else:

...     print("undetermined")

...

vaRa is more

>>>
```

Here's an example of nesting multiple "if" statements.

```
>>> ynst = 991

>>> znet = "Tssp"

>>> if znet == "Tssp":

...     if ynst > 200:
```

```
...             print("ynst out of range")

...         elif ynst < 100:

...             print("ynst too low")

...         else:

...             print("ynst within range")

... else:

...     print("out of scope")

...
```

ynst within range

As stated earlier, we can combine multiple checks inside an "if" statement using logical operators.

```
>>> ynst = 147

>>> znet = "Test"

>>> if znet == "Result" and ynst == 147:

...     print("Sweet!")

... else:

...     print("Bummer!")
```

...

Bummer!

The "znet == "Result"" check will always return false because znet is assigned the value "Tssp". The "and" operator returns true only if all operations return true which was not the case this time.

Here is another thing that makes Python stand out from other programming languages: it has no switch-case statement. Truth be told, if you properly master if-else-elif, you will never miss the switch-case statement.

Loops

Humans hate repetitive work. It's boring. However, computers excel at performing the same task over and over again. There are two major types of loops in Python: the "for" and "while" loops.

The for loop

The "for" loop is ideal if you know how many times the loop needs to be executed. Let's discuss the things you must know before working with "for" loops.

Iterable

These loops require an entity that offers multiple attributes that the loop can use to iterate over.

Iterator

The loop also requires a pointer like entity that will go over the iterable. Depending upon the iterable, the iterator stores iterable entity's data while iterating.

A simple "for" loop example is below.

```
>>> for itr in range(190,200):
...     print i
...
190
191
192
193
194
195
196
```

197

198

199

A simple program, but it highlights a few very important things.

1. range() creates a sequence of numbers starting and ending in the limits given to the method
2. The "for" loop executes as soon as the maximum limit is reached and not when it's crossed. This is evident from the program output as the loop execution stopped as soon as the value of "itr" became 100.

The range() method has a default increment of one but you can change that. But, a point to remember, only integers can be given to the range().

```
>>> for xyz in range(22,33,2):

...        print(xyz)

...

22

24

26

28
```

20

32

The range() will work even if we omit the lower limit but we can't pass a step also. For example:

```
>>> for nUm in range(7):
...     print(nUm)
...
0
1
2
3
4
5
6
7
```

An important observation here is the number started from zero which is very valuable when using range() to output indices for data structures.

Here's a question: will we get an error if we use a lower limit less than the upper limit? Let's test it out.

```
>>> for per in range(100, 50):
...     print(per)
...
>>>
```

Interestingly, no errors were given. The loop just didn't execute.

The while loop

When we do not know how many times a loop needs to iterate, we use the "while" loop. At the start of each iteration, the condition is checked and the loop block is executed if the condition is still true. This property makes it ideal to create infinite loops which are very important in various applications such as game development.

```
>>> itr = 91
>>> while (itr < 95):
...     print("Executing this statement")
...     itr = itr + 1 # can also write as itr += 1
...
```

Executing this statement

Executing this statement

Executing this statement

Executing this statement

As you can see, we have to explicitly update the "itr" variable because "while" only checks the condition. If we comment out the itr = itr + 1, we will end up with an infinite loop because value of "itr" will never become greater than 95.

"While" loops should be used with extreme care and only when you are confident with your programming skills.

Data Structures in Python

Lists

Up till now, we have seen variables that can hold one value at a time. In many cases, this is insufficient. For example, you want to store the hour marks and the task you need to do in a list.

tasklist = ['8.00','wake up', '9.00', 'work', '13.00', 'lunch', '17.00', 'off to home']

All the values in the "tasklist" are strings but we can combine different types of data in a list. The first item index of a list is always zero. Let's test that theory.

>>> tasklist[0]

'8.00'

The len() function provides the number of items in a list.

>>> len(tasklist)

8

Lists are iterables hence we can use loops to go over them. Let's use a "for" loop to access all the items in our "tasklist".

>>> for tsk in tasklist:

...   print tsk

...

8.00

wake up

9.00

work

13.00

lunch

17.00

off to home

The operator "in" is used to get one item from the list at every loop iteration. We can use an index to get a list item. Can we know the index if we have an item value? Yes we can, and here's an example.

```
>>> for tsk in tasklist:
...     print("Value:",tsk,"- Position:",tasklist.index(tsk))
...
```

Value: 8.00 - Position: 0

Value: wake up - Position: 1

Value: 9.00 - Position: 2

Value: work - Position: 3

Value: 13.00 - Position: 4

Value: lunch - Position: 5

Value: 17.00 - Position: 6

Value: off to home - Position: 7

Lists support negative values as index. For example, to get the last value of a list, we can use a negative one as index.

>>> print(implist[-1])

off to home

It is possible to replace data in a list.

>>> tasklist[2] = "Now it's 9am"

>>> print(tasklist)

['8.00','wake up', 'Now it's 9am', 'work', '13.00', 'lunch', '17.00', 'off to home']

We can verify the membership of an item in a list.

>>> if "sleep" not in implist:

...   print("sleep is for the weak")

...

sleep is for the weak

The list method of append() can be used to add new items to the list. The new item is added at the end of the list.

>>>tasklist.append("dinner")

>>> tasklist

['8.00','wake up', 'Now it's 9am', 'work', '13.00', 'lunch', '17.00',
'off to home', 'dinner']

We can use the insert() method to add an item at a specific index position.

```
>>> tasklist.insert(8,"19.00")
```

```
>>> tasklist
```

['8.00','wake up', 'Now it's 9am', 'work', '13.00', 'lunch', '17.00',
'off to home', '19.00', 'dinner']

We can remove a specific item using its value through the remove() method.

```
>>> tasklist.remove("19.00")
```

```
>>> tasklist
```

['8.00','wake up', 'Now it's 9am', 'work', '13.00', 'lunch', '17.00',
'off to home', 'dinner']

We can remove a specific item using its position with the pop() method.

```
>>> tasklist.pop(8)
```

"dinner"

```
>>> tasklist
```

['8.00','wake up', 'Now it's 9am', 'work', '13.00', 'lunch', '17.00', 'off to home']

In case no index is provided, pop() removes the last item in the list. So, in the above code we could have gotten the same result by not providing the index '8' in the pop().

The extend() method is one of the many ways to join two lists.

>>> funlist = ["weekend","24/7 fun", "no work"]

>>> tasklist.extend(funlist)

>>> tasklist

['8.00','wake up', 'Now it's 9am', 'work', '13.00', 'lunch', '17.00', 'off to home', "weekend","24/7 fun", "no work"]

All the items from "funlist" are added at the end of "tasklist".

You can nest lists as well which means a list can contain other lists.

>>> newlist = [tasklist, funlist]

Dictionaries

A dictionary is an advanced form of arrays because we can set custom index (called key) for each item (called value) in it. The key and value can be of any data type.

You can change the contents of a dictionary. A dictionary can hold another dictionary.

Let's rewrite our tasklist as a dictionary. You will immediately see the data make more sense.

```
>>> taskDict = {

... "8.00": "wake up",

... "9.00": "work",

... "13.00": "lunch",

... "17.00": "off to home"

... }
```

Instead of a numerical index, we use the key to retrieve a value.

```
>>> taskDict["9.00"]

work
```

To know how many key-value pairs are present in a dictionary, we can use the len().

```
>>> len(taskDict)

4
```

We can also get all the keys in a dictionary using the method keys() which returns a list containing all the keys.

>>> taskDict.keys()

['8.00', '9.00', '13.00', '17.00']

In a similar way, the values() method returns all the values in a dictionary as a list.

>>> taskDict.values()

['wake up', 'work', 'lunch', 'off to home']

Let's add a list to this dictionary.

>>> taskDict["Fun List"] = ["weekend","24/7 fun", "no work"]

"Fun List" is the key. Dictionaries have become ordered in Python 3.0+ that make them even faster and newly added items are always added at the end of a dictionary.

The pop() method is also supported by dictionaries.

>>> taskDict.pop("Fun List")

["weekend","24/7 fun", "no work"]

The operator "in" is applicable to dictionaries too but it works slightly different. For dictionaries, the "in" operator retrieves the keys. We can use it with a loop to retrieve the keys and values of any dictionary.

>>> for timeMarc in taskDict:

```
...        print("Time:", timeMarc,"- Task:",taskDict[timeMarc])

...
```

Time: 8.00 - Task: wake up

Time: 9.00 - Task: work

Time: 13.00 - Task: lunch

Time: 17.00 - Task: off to home

The "in" operator along with the conditional "if" statement can be used to check if certain data is present as a key in a dictionary.

```
>>> if "22.00" in taskDict:

...        print("you should be asleep by now!")

... else:

...        print("good boy")

...
```

good boy

To find something in the values, we can replace taskDict() with taskDict().values()

```
>>> if "sleep" in taskDict.values():

...        print("sleep is good!")
```

... else:

...   print("sleep is for the weak!")

...

sleep is for the weak!

We can add a dictionary inside another dictionary.

>>> newDict = {

... "a": "apple",

... "b": "ball",

... "c": "cat",

... "d": "door",

... "vowels": {'one': 'a', 'two': 'e', 'three': 'i', 'four': 'o', 'five': 'u'},

... }

Tuples

Tuples don't serve any purpose other than to provide a medium when we are required to pass information between system layers or components without the chance of change.

Here's our first tuple.

```
>>> tasktup = ('8.00','wake up', '9.00', 'work', '13.00', 'lunch', '17.00', 'off to home')

>>> print(tasktup)

('8.00','wake up', '9.00', 'work', '13.00', 'lunch', '17.00', 'off to home')
```

Python allows you to change the type of a variable. For example str(64) will convert it into '64'. We can use the same concept to change a tuple. We first convert the tuple to a list, make the edits, and convert the updated list back to a tuple.

```
>>> templist = list(tasktup )

>>> templist.pop()

'off to home'

>>> templist.insert(8,"overtime")

>>> tasktupnew = tuple(templist)

>>> tasktupnew

('8.00','wake up', '9.00', 'work', '13.00', 'lunch', '17.00', 'overtime')
```

Sets

Python has maintained the rules of mathematical set theory when creating its "sets" data structure.

>>> taskset = set(['8.00','wake up', '9.00', 'work', '13.00', 'lunch', '17.00', 'off to home'])

Since the sets are unordered, there's no index. Therefore, you cannot change items in a set using an index value, but you can add new items or remove existing ones.

We can use all the other methods we have learned so far and also some more such as add() and update(). Both add new elements at the end of the set. You are already familiar with the remove() and discard() and both are supported by sets.

Chapter 3: Shifting Gears

Python is very extendable, thanks to the hundreds of external libraries written by volunteer contributors. In this chapter, we are going to work with some of the external libraries, how to install and import them, and use the methods available in those libraries.

External Libraries

We are going to use the popular chart creation "matplotlib" library as a means of demonstrating how to install and import an external library.

We can use pip to install matplotlib. Open the Windows command prompt and run the following command. A note here: If you are using an advanced interpreter like PyCharm, the command prompt is accessible from the interpreter. You do not have to open command prompt separately.

pip install matplotlib

Python will download and install the library and all the required resources. Once it is setup, we can import the library to start using it.

import matplotlib.pyplot as plt

When we ask Python to import a library or your own written module, there is a path registry that Python searches to find that resource. The following code will give you all the paths registered in the registry.

```
>>> import sys

>>> for resourcePath in sys.path:

...     print(resourcePath)

...
```

You can import your own script in another script. Place your script in one of the paths the above code outputs and you will be able to use "import" to use it in another script.

The import process takes a lot of resources and can make your interpreter or entire computer become unresponsive. Do not worry, just give it a minute, the process will finish and everything will be back to normal. If you are running in interactive mode and there's no output, it means the import was a success. Otherwise, there will be an error.

Because the import process takes too many resources, it's only run once for each resource. Duplicate import for the same library/module is ignored.

```
import datetime

import datetime
```

In the above two import statements, only the first one will run. Python will not execute the second statement. It means you cannot recursively call your module just like you can recursively call a function to do something. If the module/library has various functions (methods), we can import a specific one just like we did with the matplotlib.pyplot. Once imported, we can start using it.

>>> plt.plot([10, 13, 15, 17])

[<matplotlib.lines.Line2D object at 0x000001905095FF88>]

The pyplot is a simple line chart or line graph. If you look closely, we have given coordinates using the plot() function.

>>> plt.ylabel("Y Coordinates")

Text(0,0.5,u'Y Coordinates')

>>> plt.xlabel("X Coordinates")

Text(0.5,0,u'X Coordinates')

>>> plt.style.use('grayscale')

The "Text" outputs are just a confirmation that the axes labels are updated to the new values. To view the plot, we use the show() method.

>>> plt.show()

The plot is shown in a new window. It looks like below.

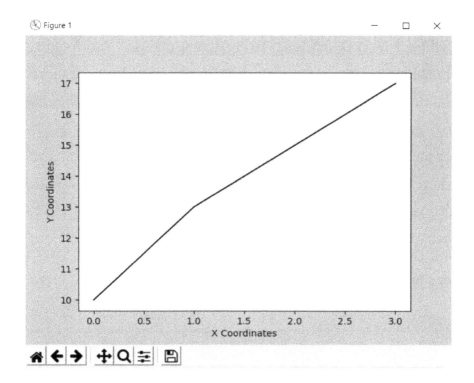

We set the plot style as grayscale otherwise the line would have been colored. There are different tools available to edit or save this plot. The plot remains open until you close it using the close button. This is important because the interactive mode might be unresponsive until you close the graph.

The matplotlib library is a huge resource to visualize data in many different ways. It's not possible to cover it completely in this book. We used it as an example for importing and using an external library. Later on, we will use another library to create graphs and plots.

Strings

Python supports extreme string manipulation. The language considers strings as a data structure, specifically a list of characters. This enables the programmer to manipulate strings in ways not possible in other programming languages. The advanced methods make Python also better for data analysis where the data is string most of the time. Data type conversion is also very straightforward. Python handles all string operations faster than the competition.

```
>>> op = "974"

>>> type(op)

<class 'str'>

>>> pp = 'this is string'

>>> type(op)

<class 'str'>
```

Enclose anything in single or double quotes for Python to consider it as string type. Notice how Python returned "class". This has changed in recent Python versions so the language has become more uniform with the statement "Everything in Python is a class."

```
>>> strJ = op + pp
```

>>> print(strJ)

974this is string

The '+' operator acts as a concatenator for strings. We can also use the join() method to concatenate strings.

>>> strJ = op.join(pp)

>>> print(strJ)

t974h974i974s974 974i974s974 974s974t974r974i974n974g

The output is a little weird and unexpected but this is just how join() method works. The result looks more like a multiplication of both strings. The join() method breaks the argument string into individual elements and joins the string the method is applied to with those individual elements.

>>> str(00457)

  File "<stdin>", line 1

    xx = 00457

          ^

SyntaxError: invalid token

In newer Python versions, we cannot put leading zeros, it gives a syntax error. There's a work around though: the zfill() method.

```
>>> str(457).zfill(6)
```

```
'000457'
```

The zfill() method is only applicable on strings, so we need to convert the number to string before adding the leading zeroes. There are already three digits in 457 so we gave six as zfill() argument because we want to add three leading zeros making the total characters in the string six.

The len() works on the string the same way it works on other data structures - it gives the number of elements.

```
>>> exStr = "139 characters"
```

```
>>> len(exStr)
```

```
13
```

Combine len() with range() and we can dissect the string to access each character individually.

```
>>> for chr in range(len(exStr)):
```

```
...     print(exStr[chr], end=' ')
```

```
...
```

```
1 3 9   c h a r a c t e r s
```

Some observations.

1. Just like lists, we can access the string using its index
2. To stop Python from adding a newline after the print output, we have added a comma ',' which adds a space instead of a newline

There are several ways to access the elements of a string. Let's try another method that uses the 'in' operator.

>>> for chr in exStr:

...   print(chr)

...

139 characters

Fewer characters to get the exact same output. The first character has the index zero, as proven by the example below.

>>> exStr[0]

'1'

The last element has an index which is one less than the length of the string.

>>> exStr[len(exStr)-1]

's'

All this information can be used to utilize a while() loop.

```
>>> ind = 0

>>> while ind <= len(exStr) - 1:

...      print(exStr[ind])

...      ind = ind + 1

...
```

139 characters

In the while loop, we could have wrapped "len(exStr) - 1" within parentheses but it isn't required for our current example.

Slice and Dice Strings

Slicing a string is very easy using the indices. Slicing creates a new string from the string.

string[start index:optional stop index:optional step amount]

Let us have this slicing technique applied to the exStr string.

```
>>> exStr[4:]
```

'haracters'

Since we did not provide a stopping index, the returned string has all the characters from the starting index to the end of the original string. The easiest way to reverse a string is to use slicing without the start and stop indices and use a step value of negative one.

```
>>> longstr[::-1]
```

```
'sretcarahc 31'
```

Extraction and Replacement of Substrings

The creators of Python have tried their best to have Python make sense in every way possible. The method names almost always closely resemble their application. Let's focus on the string method split(), which splits the string according to the given substring.

```
>>> exStr.split()
```

```
['13', 'characters']
```

Since we didn't provide any argument to split(), the default value of space ' ' was used. This is apparent from the example below.

```
>>> 'kajhskjdha'.split()
```

```
['kajhskjdha']
```

Let's provide a substring.

```
>>> 'kajhskjdha'.split('a')
```

```
['k', 'jhskjdh', '']
```

The returned list of strings do not have the argument we used to split the string. We can also replace a portion of a string with something else.

>>> exStr.replace("13","some")

'some characters'

Note that replace() doesn't change the real string. We can definitely check the contents of exStr.

>>> exStr

'13 characters'

The replace() by default replaces all the instances of a specific substring with the required substring.

>>> exStr.replace('a','e', 1)

'13 cheracters'

Not just a single character, replace can be used to replace entire words or phrases with something new. You might have used the find and replace tool on many software applications that deal with text such as word processors. Notice how easy it is to implement such as feature using the replace() method. Maybe a future challenge could be to develop a word processing application using Python?

Exceptional Handling

Error Types

There are two main categories of errors in Python.

Syntax errors

When you do not follow the rules of script writing in Python, you will get a syntax error, also called parsing error. We have already seen an example of syntax/parsing error.

>>> str(00457)

  File "<stdin>", line 1

   xx = 00457

        ^

SyntaxError: invalid token

Whenever an error occurs, Python tries to provide an explanation so the programmer knows what needs to be fixed. The explanation and hints are sometimes not very easy to understand but most of the time, you should consider the hints carefully.

Exceptions

If you make no mistakes in syntax during the program writing, you might still end up having the wrong output or breaking program flow with an error. Let us discuss some more errors.

Index error

Using an index that is not present in a data structure will lead to an index error.

```
>>> listrand = ["python","the","best", "programming", "language"]

>>> listrand[100]

IndexError: list index out of range
```

Name error

If we use a resource without assigning a value to it first, we get an error.

```
>>> listrand = ["python","the","best", "programming", "language"]

>>> print(listrandom)

NameError: name 'listrandom' is not defined
```

Type error

When we try to perform an operation that is not supported by the given data, it leads to a type error.

>>> intpp = 741

>>> strpp = "007"

>>> strpp + intpp

TypeError: can only concatenate str (not "int") to str

If you apply the arithmetic operator '+' but the position operands is reversed, the explanation of type error becomes different.

>>> intpp + strpp

TypeError: unsupported operand type(s) for +: 'int' and 'str'

Import error

If you import a library/module that Python is not able to find in the directories listed in the path directory, you will get an import error.

>>> import custmod

ModuleNotFoundError: No module named 'custmod'

>>> from math import triag

ImportError: cannot import name 'triag' from 'math' (unknown location)

We see two different descriptions for the same error because the error happened due to different reasons.

Logical error

When you make an error that's not an error for Python but we still don't get the output we were expecting, then we have encountered a logical error. Logical errors are very difficult to find and fix because you do not get any hint as to what's causing it.

Let's write a script with a logical error to illustrate how a logical error can occur. Even with this simple example, it would be a serious task to find the cause of error because it's really not apparent.

import math

num1 = 200

num2 = num1 / 5 * math.pi

print num2

The following output is generated after saving and executing the above script.

125.66370614359172

The result is wrong because we were expecting an output of 12.732395447351628. We know that there is something wrong with the program, but we have no idea where the issue is. We will have to deploy troubleshooting techniques to find the problem. We know that Python always follows PEMDAS when calculating the result of an expression containing arithmetic operators. Since multiplication and division have the same priority, Python is going from left to right with the execution. The "num1" is divided by "5" and that result is multiplied by the pi value. We have to use brackets to force Python to executing the problem.

num2 = num1 / (5 * math.pi)

The correct output will be shown now.

12.732395447351628

Exceptional Handling

All errors, except logical ones, break the execution flow. This is unacceptable if you are creating an application. In earlier days, it used to be very common that an unexpected situation would break the application or system. Thanks to better programming practices now, exceptional handling is now part of all high-level programming languages and all programmers utilize it to create distributable applications. What is exception handling? The technique to handle errors so they don't break the program execution. For example, we know users will enter bad data even though we have given all the instructions. It is always a good idea to wrap the input() statement inside a try-except pair. Even if the data provided by the user is correct, it is very easy to forget that input() takes the input as string. Applying any method or operation on this input that is not compatible with strings will result in an error.

```
>>> try:
...     capT = input('Enter a number: ')
...     res = capT / 2
...     print(res)
... except:
...     print("There is an error in the try block")
...
There is an error in the try block
```

We tried to divide a string and strings do not support division. An error has occurred but because it's in a code that's wrapped with "try", execution flow doesn't break. However, the entire code block is ignored and the code block wrapped with "except" is executed. Hence, the output that we see is "There is an error in the try block".

There is also the "else" wrapper that we can use to execute code block in case the "try" block code doesn't raise an exception. Remember that if there's no error, the "try" also gets executed. Here's an example proof.

```
>>> centRy = 113

>>> try:

...     print(centRy)

... except:

...     print("No exceptions this time")

... else:

...     print(centRy)

...

113

113
```

There's another option, "finally", code wrapped with it which will be executed no matter what the result of the code block inside "try" was.

```
>>> chk = "This value is safely stored."

>>> try:
```

```
...        print(chk)

... except:

...        print("Something is wrong")

... finally:

...        print("I will persevere")

...
```

This value is safely stored.

I will persevere.

Best Troubleshooting Practices

In the above section, we saw the simplest possible examples for each type of error. Sometimes even with the Python hint, we do not know what the cause of the error is or how to fix it. This happens because the explanation provided by Python doesn't make sense to us. Let's see our previous example again.

```
>>> str(00457)

  File "<stdin>", line 1

  xx = 00457

      ^
```

SyntaxError: invalid token

We know we have made a mistake, and we also know there is a syntax error. But, we don't know that Python doesn't support leading zeros and that's what is causing the error. "Invalid token" doesn't make sense unless you already know what it means. If you are seeing this error for the first time, you will have to research to understand what has happened and what needs to be changed. As stated before, logical errors are even harder to fix because we have no idea what and where things went wrong. Imagine a script of thousands of lines of code and you are not getting the expected output. What to do next? Where do you start? How do you track the issue smartly so you do not waste more time in troubleshooting than you did coding? There are some best practices when it comes to debugging and troubleshooting.

Planning before execution

There is a reason companies spend more budget and time at planning a particular business venture than actually doing it. All the aspects are carefully analyzed and discussed to remove any loopholes. There are various tools that help here. In programming, we have algorithms (textual) and flowcharts (visual). Both tools should be deployed, but algorithms are usually used before flowcharts because algorithms are system independent. They are used to roughly estimate the process. Flowcharts are much more detailed and easier to understand as they not only show the flow of execution but also the different requirements.

Leave breadcrumbs

Programming is like exploring a huge cave or walking a labyrinth. If you did not leave breadcrumbs to help you back track, you will be lost forever! Comments are useful because most of the time there are multiple people working on the same project and everyone needs to understand what you have written. It also helps you remember what you did when you were coding because, believe me, no one remembers what they did. If you don't remember how you resolved a problem or why you coded a specific way, it's very difficult to update it or correct it.

Dry run

A lost art - this used to the the main troubleshooting method. There are now better tools such as live trackers that you can utilize to see all the changes happening in a program. But, this tradition of using the pen and paper and doing it the hard way has lead to great discoveries. I still prefer that method. There's just something about it that helps me visualize the links between different things and find the problem. Of course, that doesn't work in a script that has thousands of line. For those projects, you definitely need the latest debugging tools.

Test unexpected behaviour

Programmers are problem solvers. They think in a specific mindset but that might not always be how the end user would think or use the product. The best way to make sure your program does not break because the user entered "one" instead of '1' is to unit test the script with millions of different inputs and analyze the result. If the script breaks in certain situations, you can program to fix them.

Debugging in Python

The "pdb" tool is the standard debugging tool in Python and works the same way debuggers work on other platforms. You can add a marker so the debugger would start from there and then execute your code step by step, monitoring all the changes your script makes. Here's a script that we are going to examine with the "pdb" debugger.

```
import pdb


pdb.set_trace()

num = input("Enter a number: ")

try:

        numAct = inp(num)
```

except:

```
    print(type(num))
print('Multiplication table of %d', %num)
for mult in range(1, 21):
    print("%d x %d = %d" % (num, mult, num*mult))
```

A simple multiplication table generator script where the first line has the "pdb.set_trace()" set. Before we could use the tracer, we have to import the "pdb" library. The debugger will start from the very first line. A small window will appear that belongs to the debugger and can take various commands. You can start debugging by passing many commands including the following.

1. (s)tep - execute the line of code debugger is currently positioned at
2. (c)ontinue - execute the remaining code unless a breakpoint is met, in that case stop the execution there
3. (n)ext - works well on functions and executes until a newline is met or until a return is encountered

Use the commands to see how your program changes with the execution of each line.

Functions

Python supports object-oriented programming as well as functional programming. In the latter, programmers can name some lines of code and use that name to run those lines of code as many times as needed. The lines of code defined by a name are called functions. You can also make the function take on arguments which can be passed when calling the function.

```
>>> def prntseq(num):

...      # print sequence of numbers from 1 up till the given number

...      cnt = 1

...      while (cnt <= num):

...              print(cnt, end=", ")

...              cnt = cnt + 1 # can also write as cnt += 1

...
```

We have created a simple function using the "def" keyword. The function, named "prntseq" prints a sequence of numbers from one to the ending limit passed during the function call. The "num" function variable takes on the passing argument so we can use it in the function *definition*. Let's run the function with an argument of 10.

>>> prntseq(11)

1, 2, 3, 4, 5, 6, 7, 8, 9, 10, 11

We can also give an expression as an argument and Python will pass the result to the function. Here is a simple example.

>>> prntseq(5*3)

1, 2, 3, 4, 5, 6, 7, 8, 9, 10, 11, 12, 13, 14, 15

A function can take an unlimited number of arguments, but it's good practice to limit arguments of every function to a maximum of two. We should divide our function into smaller functions if there is a need to use more than two arguments. We created a multiplication generator script a couple of pages ago. Let's create a function to get the same output.

>>> def mlt_tbl(mul, timz):

...   print('Multiplication table of %d' % mul)

...   for seq in range(1,timz+1):

...   print("%d x %d = %d" % (mul, timz, mul*timz))

...

>>> mlt_tbl(9,12)

Multiplication table of 9

9 x 1 = 9

9 x 2 = 18

9 x 3 = 27

9 x 4 = 36

9 x 5 = 45

9 x 6 = 54

9 x 7 = 63

9 x 8 = 72

9 x 9 = 81

9 x 10 = 90

9 x 11 = 99

9 x 12 = 108

When there are hundreds of functions in your script there are few things to keep in mind.

1. Add exception handling whenever using arguments in expressions because the wrong data type can break execution
2. Create a function template and stick to it for all functions. This will help you have better expectations. A function

definition should be divided into smaller chunks such as imports, variable assignments including all input(), actual work on the data, and in the end the return or other form of output such as print().

3. Use operators and methods that return the same output for all or most of the data types.

>>> def dosum(numx, numy):

...   print numx+numy

...

>>> dosum(7845, "this a string")

TypeError: unsupported operand type(s) for +: 'int' and 'str'

This shows how important it is to deploy exception handling and type conversion together when creating functions.

The Concept of Recursion

You can call a function inside its own definition creating a sort of loop to perform the same action consecutively on a data. It's like using a hammer to put up a nail on the wall. A lot of times, especially when dealing with numbers, we perform an operation that gives us a result and we have to perform the operation on that result again. Here's a simple example of recursion to find the factorial of an integer. Factorial of a number is the product of all numbers from one to that number. For example, the factorial of 4 is 1*2*3*4 = 24.

```
def get_fact(num):

    if x == 1:

        return 1

    else:

        return (num * get_fact(num - 1))
```

When we test the above function with "print(get_fact(6))", we get the output as 720, which is correct because 1*2*3*4*5*6 = 720. Note that we have declared a base condition, a condition that once met will stop the recursion. In the above code, "x == 1" is that base condition. We are using the knowledge that when a function is called from another function, the execution of the latter function is suspended until execution of the former function ends.

Variable Scope

In Python, variables have local scope by default. What does local scope mean? It means if you define a variable inside a function definition, it is not accessible in another function's definition. If we want a variable to be available in all function definitions, we can declare them outside of all function definitions. Another method is to use the "global" keyword.

```
x = 123 # this is a global variable


def tstFun():

    y = 789 # this is a global variable

    print(x + y)


tstFun()
```

Class and Objects

When we build something in real-life, we need a blueprint or schematic so we know what to put where. Classes are the blueprints of programming languages that are used to construct objects. They are the basis of object-oriented programming and help program complex concepts by replicating the same features multiple times.

```
>>> class frstcls:

...     prp = 554

...
```

Variable "prp" is known as a class property. We can create new objects once the class has been defined.

>>> obj1 = frstcls()

>>> obj1.prp

554

Every object created using the frstcls() class will inherit the property "prp". We can also create functions to the class that will become object methods. In fact, every class must have the __init__() function which is sort of an initiation function. The __init__() function is used to declare all properties of the class and optionally set their values, it can also be used to declare dependencies. Classes and objects are much more prevalent in video games where these concepts are used to create non-player characters to enrich the gaming experience. Most non-player characters share the same design and mechanics with cosmetic changes.

>>> class npc:

...   def __init__(instance,type,id,health,armor,mechanics):

...   instance.type = type

...   instance.id = id

...   instance.health = health

```
...            instance.armor = armor

...            instance.mechanics = mechanics

...
```

Every object created by the class is its "instance" which is what is passed to the __init__() function. Let's use our class to create an npc player.

```
>>> npc1 = npc("Mercenary", "Sailor", 75, 0, [2, 'walk', 'shoot'])
```

We can access individual object parameters easily like below.

```
>>> npc1.armor
```

```
25
```

```
>>> npc100 = npc("Level Boss","Pirate King",1000,500,[5, 'fast move', 'double 'jump', 'move tracker', 'slasher', 'unlit revolver'])
```

If you think that's a tough one, wait for the final level boss!

All object parameters are mutable unless we have an immutable data structure like a tuple.

```
>>> plyratk = 80
```

```
>>> if plyratk >= 75:
```

```
...     npc1.hp = 0
```

... else:

...   npc1.hp = npc1.hp - plyratk

...

>>> npc1.hp

0

We can perform all other methods that we have used for data structures and variables up till now, and then some on objects.

>>> del npc1.specialmoves

>>> del npc2

Scheduling and Automating Python Scripts

We can use the "Task Scheduler" that comes standard with all Windows installations to schedule and automate any tasks including execution of Python scripts.

Search "Task Scheduler" in Windows 10 start menu. An application with the same name will appear in the search results, select it. Currently, the application looks like the following screenshot.

A new window "Create Basic Task Wizard" will popup when "Create Basic Task" is clicked.

The step by step process is detailed below:

1. Enter a memorable and recognizable name so identifying different tasks later would become easier. It is also recommended to add a concise description. Click "Next".

2. We have to set a trigger which can be a specific time or an event in another Windows application or system. Whenever the trigger would occur, Windows will run the scheduled task. Select "Daily" for now.

3. Let us set 11:59pm "everyday" and whatever the current date is as the starting point. Keep one in the recur field because we are going for a daily schedule. Click "Next".

4. We want to run a Python script so we will keep the option "Start a program" selected. Click "Next".

5. In the final step, we have to direct Windows which program we want to schedule. To schedule a Python script, we should first run Python interpreter and then load the script as an argument. For example, if the

Python interpreter is located at "C:\Python\python.exe" and the script you want to run is located at "D:\myscript.py".

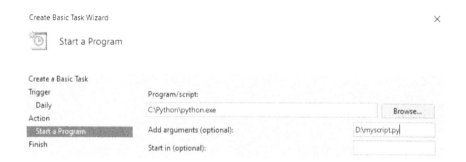

Click "Next" and you will see a summary of the task you have setup. Click on "Finish" to create the scheduled task you have setup that will run everyday at 11:59 pm.

It is such a good way to automate your daily tasks!

If you do not want to pass an argument or if the above process does not work for you, you can also schedule a Python script through the Windows Batch (.bat) file that are basically Windows script files. To schedule this way, we can add the following instruction to a batch file.

C:\Python\python.exe "D:\myscript.py"

Make sure to save the file. Now, we can schedule the batch file using the Windows "Task Scheduler" and Windows will execute the Python script.

This was a tutorial on how to create a very simple scheduled task. The "Task Scheduler" also offers many advanced options that are out of the scope of this book but you should explore them. There are a lot of similar tools available in Windows that an average user does not know about. It's always fun to explore new things!

Chapter 4: Basic Data Analysis

Everything in this world is constantly generating data. If we can record that data, we can process it to find the hidden meanings. It can help in making important decisions regarding your business or life. In this chapter, we will take a look at the basic data analysis concepts.

Data Parsing

There are many different ways to access data from the same file type. There are several libraries and depending upon the Python installation for the operating system in use (Windows in our case), there are even more options available.

Excel

MS Excel is the most widely used spreadsheet software in both personal and commercial domain. Due to this reason, there are several options to get data from MS Excel spreadsheets. In the below example, we are leveraging the Windows client for Python to get data from an excel sheet. Note that this Windows client method can be used to integrate with any application running on Windows operating system.

import win32com.client as win32

```
exc=win32.gencache.EnsureDispatch('Excel.Application')
```

```
exc.Visible=True
```

```
filepath = "" # set the file path here
```

```
wrkbk = exc.Workbooks.Open(filepath)
```

```
wrksht = wrkbk.Worksheets('Sheet1') # get the sheet with the
```
name "Sheet1"

```
wrksht.Name = 'Python Created' # change the name of that sheet
```

```
wrksht.Cells(1, 1).Value = "Hello" # set the value of the first cell
```
in the worksheet

As you can notice, the first cell in the worksheet has row and column values of one.

An alternate approach

An alternative is to use another library specifically made to read data from an Excel file. We are going to use "xlrd" library.

```
wrkbk = xlrd.open_workbook(filepath)
```

One issue with "xlrd", which is not an issue but an inconvenience when you are developing a script, and actually an improvement from the user-side perspective, is that the Excel file is not visible so you have no idea what the Excel file looks like.

Once we have the workbook, we can access any sheet using its name or index.

wrksht = workbook.sheet_by_name('Sheet1') # access the sheet by the title of 'Sheet1'

wrksht = workbook.sheet_by_index(0) # access sheet with its position

In an Excel workbook, the position of a sheet can be changed so it's better to get sheets using their name and not the index.

print wrksht.cell(0, 0).value

In the "xlrd" library, the first cell of a sheet has a row and column indices of zero. To get all the values in a spreadsheet, we can use nested for loop to access each sheet and all the cells in it. One problem though, if we accidentally try to access a cell that is empty, we will get an error. For example, consider that the cell(100, 100) is empty. The following will result in an error.

print wrksht.cell(100, 100).value

To avoid this issue, we can read the number of rows and columns with data in an Excel sheet like below.

print("number of rows:", wrksht.nrows) # get number of rows with data

print("number of columns:", wrksht.ncols) # get number of columns with data

Using the above information, we can correctly set the limits on our "for" loops.

CSV

Comma Separated Values (CSV) files are a common method to distribute data over the internet due to its low file size. In data analysis, most of the times you will be reading and writing data in this format. They are essentially a text file where every data value is separated by a comma. For example, here is an example of some CSV data.

name,department,designation

Sinclair,IT,network engineer

Jones,Sales,inside salesperson

Sasha,HR,hiring manager

Alexa,Admin,VP secretary

The code to read the comma separated data from a CSV file saved in the same directory as the script file.

import csv

import os, sys

```python
curdir=os.path.dirname(sys.argv[0]) # gets the directory where
the script file is located

filep = os.path.join(curdir,"employee_data.csv")


with open(filep, mode='r') as csv_file:

    csv_read = csv.DictReader(csv_file)

    line_cnt = 0

    for row in csv_read:

        if line_cnt == 0:

            print('Column names are',row.keys())

            line_cnt += 1

        print(row['name'],"works                    in ",row['department'],"as a/an ",row['designation'])

        line_cnt += 1

    print('Processed',line_cnt,"lines")
```

This will output:

>>> Column names are ['name', 'department', 'designation']

Sinclair works in IT as a/an network engineer

Jones works in Sales as a/an inside salesperson

Sasha works in HR as a/an hiring manager

Alexa works in Admin as a/an VP secretary

Processed 5 lines

The first line of output that has the column names might not have the column names in this order because we are reading data by storing in a dictionary and we already know dictionaries are unordered data structures. When reading as a dictionary, the key is set as the first row header value of that column and the data from the relevant row. But, what would happen if there are no column headers? We might have to add a column header before we use this method.

To write a csv file, we can store all our data in a dictionary and that dictionary content is written to a csv formatted text file.

```
import csv

import os, sys


curdir=os.path.dirname(sys.argv[0])

filep = os.path.join(curdir,"join_date.csv")
```

```python
with open(filep, mode='w') as csv_write:

    col_hdr = ['Name', 'Join Date', 'Salary']

    fin_file = csv.DictWriter(csv_write, fieldnames=col_hdr)


    fin_file.writeheader()

    fin_file.writerow({'Name': 'Sinclair', 'Join Date': '12-Mar-13', 'Salary': '45,000'})

    fin_file.writerow({'Name': 'Jones', 'Join Date': '07-Jun-09', 'Salary': '58,500'})

    fin_file.writerow({'Name': 'Sasha', 'Join Date': '07-Jun-09', 'Salary': '85,000'})

    fin_file.writerow({'Name': 'Alexa', 'Join Date': '07-Jun-09', 'Salary': '65,500'})
```

Websites

Everything you do online is recorded by someone with the chief purpose of knowing about you and marketing you relevant ads. It sounds cool but, personally, I think it pushes humans to become too much of a habitual animal. If you keep getting bombarded with things you like, not only its frustrating, it also affects your behaviour. For example, I like sushi but that doesn't mean wherever I go I only want to eat sushi! Where's the fun of exploring new food and places and experiencing the adventure of trying the unknown? I do not want everything I do recorded and analyzed. Or, if I do, I want to be compensated for it!

But, truth be told, there are various other applications of data recording and analysis besides shoving ads in your face. There have been huge strides in technology, especially in the fields of medicine and agriculture by knowing more about the environment and all the other factors involved. Precise predictions are regularly made about the weather which help the general public in planning different aspects of their daily lives.

The world wide web is a huge resource of free and paid data. You can tap into this resource to create applications to make lives better for the people. But, the rules of ethics should not be broken in any case. It's the weight of the world on the shoulders of data analysts and scientists - if you stumble upon information that is true but too dangerous, do you show it to the world or forward it to the global stakeholders? As Spiderman once said, power comes with responsibility!

Python has several libraries for web scraping as well. We are going to use "scrapy" now. Note that "scrapy" is usually used for large projects. For simpler web scraping projects, we can use libraries such as "BeautifulSoup".

Using Scrapy

You have to install the "scrapy" library before you use it.

pip install scrapy

After the installation is finished, we have to setup a project. This is done for every website we have to scrap. In the "scrapy" library, each scraper we build is called a spider, because spiders crawl. Using the Windows command prompt, navigate to the directory where we want to create the project and then run the following command.

scrapy startproject frstspydr

You do not have to name your crawler spider, especially if they give you the heebie jeebies! A new folder with the name "frstspydr" will be created in the directory that we navigated to.

The first step is to update the "\frstspydr\frstspydr\settings.py". In the "settings.py" file, two changes need to be made. We have to uncomment the "ITEMS_PIPELINE" tuple in case it's commented out and make edits so the file content looks as below.

#Export as CSV Feed

FEED_FORMAT = "csv"

FEED_URI = "reddit.csv"

The above lines indicate we are going to save the scraped information in a CSV formatted life. We are using a Reddit name because we are going to scrap a Reddit thread.

Going back to the command prompt, run the following command.

scrapy genspider redditcrawl www.reddit.com/r/breakingbad/

The above command will create a new spider inside the "\frstspydr\frstspydr\spiders\" folder using a standard template. We are going to use this newly created spider to get all the text and images from theBreaking Bad show Reddit thread.

If you open the "\frstspydr\frstspydr\spiders\redditcrawl.py" file, the below code is visible.

```
import scrapy


class RedditbotSpider(scrapy.Spider):

                name = 'redditcrawl'

                allowed_domains =
['www.reddit.com/r/breakingbad']

                start_urls =
['http://www.reddit.com/r/breakingbad']
```

```
def parse(self, response):

    pass
```

Some observations:

1. The spider can be named anything
2. The allowed_domains adds security by setting the domain that can be crawled using this crawler. This is entirely optional
3. Upon successful crawl, the parse function is called and executed

Next step is to update the start_urls argument in the RedditcrawlSpider class so it has the correct HTTPS format URL. Most websites nowadays have the secured HTTPS and we cannot access data from those sites using the unsecured URL version.

```
start_urls = ['https://www.reddit.com/r/breakingbad/']
```

See the "pass" inside the parse() function. Replace that with the following code block.

```
def parse(self, response):

        # Extract the html text with the css class
selector

        title =
response.css('h3._eYtD2XCVieq6emjKBH3m::text').extract()
```

```python
        images_url =
response.css("img::attr(src)").extract()

        # Gather the data for every row

        for item in zip(titles,images_url):

            # store all the data in a dictionary

            scraped_info = {

                'title' : item[0],

                'images_url': item[1],

            }


        # Send all the scraped info to scrapy

        yield scraped_info
```

We have used the html class selector. We are going to use the css class values of html to get the text content (the post titles) and featured image URLs. Let's save the file and start using the crawler. The following command starts the crawler bot.

scrapy crawl redditcrawl

Before running the scrapper, make sure to change the working directory using the command prompt to the directory where the redditcrawl scraper resides. Not changing the directory and running the above command will result in an error.

While the crawler is running, it will output a lot of lines on the screen, but you can ignore all of them. After the crawl process is finished, you can navigate to the folder where you saved the crawler and find the reddit.csv file. The file will have all the post titles and direct URLs to the featured images, one in a row.

We can also download the images by creating a pipeline

```
ITEM_PIPELINES = {

 'scrapy.pipelines.images.ImagesPipeline': 1

}

IMAGES_STORE = '' # add the folder path where the images should be stored
```

If you are going to download the image files, make sure to comment out the CSV creation code we added earlier because sometimes performing both together leads to conflicts.

You can customize the crawler to get anything from any website. The "scrapy" framework supports multithreading which means you can deal with tons of online data without any compromise on performance. Still, be careful because this framework uses a lot of processing power. Smart programming goes a long way here.

Using BeautifulSoup

Beautiful Soup is a web scraping library in Python better suited for simpler tasks. It is way faster and easier to use to extract specific information from a website. First we install the library.

pip install BeautifulSoup4

Beautiful Soup extracts data from a website but is not able to make a connection with it. We need another library such as "urllib.request" to establish a secure connection with the website so Beautiful Soup can work.

We will scrape data from "example.webscraping.com" which has 25 pages containing the names and flags of all the countries in the world. The homepage URL is:

http://example.webscraping.com/

The URl of the last page is:

http://example.webscraping.com/places/default/index/25

This website is unsecured and the URLs share a common structure. Utilizing the inspection tool of Google Chrome, we inspect the html markup of the website. We notice that individual html elements do not have unique attributes, therefore we will extract accordingly by filtering through specific html elements.

The complete code is presented below along with comments for easier understanding.

```python
from bs4 import BeautifulSoup

import time

import urllib.request


webURL = "http://example.webscraping.com/"


for pag in range(0, 26): # 0-26 because we have 25 pages where second page is '1'

    if pag == 0:

        lnk = webURL

    else:

        lnk = webURL + "places/default/index/" + str(pag)

    try:

        time.sleep(1)
```

""

Beautiful Soup reads data very quickly, too quickly for most servers' liking. If they see your script sending several requests to the site very quickly, they will think you are trying to hack or DDoS the site. To prevent that from happening, the server can temporarily block your IP. We are adding a delay of 1 second between consecutive server requests so that situation never happens.

'''

```python
        page = urllib.request.urlopen(lnk)

        soup = BeautifulSoup(page, 'html.parser')

        cntry_names = soup.find_all('td') # get all
country names in a page

        cntry_flags = soup.find_all('img') #get all flag's
image URL in a page

    except:

        print("Something bad happened while trying to
scrap the website.")

    if (cntry_names): # only run the following 'for' loop if
content_id is not empty

        for  name in cntry_names:
```

```python
            print(name.text.lstrip()) #lstrip() removes
```
any unwanted spaces from the text

```python
        if (cntry_flags): # only run the following 'for' loop if
```
content_img is not empty

```python
            for itr in cntry_flags:
```

```python
                print(webURL + itr['src'].lstrip('/'))
```
#lstrip() removes extra forward slashes from the URL

The above code opens each website page and gets the country name and the country flag. We can also store this gathered data in a CSV file so we can access it later without having to scrap the website another time.

```python
f = open('scrapData.csv','w')
```

```python
if (content_td): # only run the following 'for' loop if content_id
```
is not empty

```python
        for itm in content_td:
```

```python
            f.write(itm.text.lstrip())
```

```python
if (content_img): # only run the following 'for' loop if
```
content_img is not empty

```python
        for itr in content_img:
```

```python
            f.write(webURL + itr['src'].lstrip('/'))
```

Working with NumPy

The "numpy" library provides us with one more data structure: the numpy arrays. These arrays closely resemble the mathematical matrices. In fact, all the matrix concepts can be applied on the numpy arrays. We have to install the array before using it.

pip install numpy

Once installed, we can start using the library.

import numpy as np

arr = np.array([36, 69])

brr = np.array([[75, 53],

[95, 51]])

print(arr.ndim) #get dimension of array which is returns the number of rows in the array

The number of elements in each row must be the same in a multidimensional array. The output of the above script is:

1

To get the number of rows and columns, we have to use the "shape" property.

print(arr.shape) # for a multidimensional array, returns a tuple containing number of rows then number of columns, for a single dimensional array, the number of columns is returned

print(brr.shape)

The output is:

(2,)

(2, 2)

The numpy arrays look very similar to nested lists, but support many methods that we cannot apply on nested lists. To see how many items are in an array, we use the "size" property.

print(brr.size) # returns 6 because total 6 elements are present in the "brr" array

Individual elements in the array can be accessed using the indices just like in a list, but the syntax is a little different. In a numpy array, the first row - first column position has 0, 0 index.

print(arr[1,1]) #outputs 53 because 1, 1 means second row and second column

As the name suggests, *num*py arrays are geared towards number types. But, we can also store strings with the requirement that all elements in an array must be of same data type.

crr = np.array([111, "is called", " a Nelson", True])

print(crr)

The output will be.

['111' 'is called' ' a Nelson' 'True']

The numpy library converted all the data to the string type. We have to be careful because during data analysis, if we program our script to find the boolean "True" value in this array, the script will always return null because the boolean "True" has been converted to the string "True". Another observation is that the commas are missing but that's just what printing a numpy array looks like. We can print another array just to be sure.

print(bArr)

[[75 53]

 [95 51]]

What if we have an array that has only numbers and Boolean values? Here is an example:

drr = np.array([183, True, False])

print(drr)

The output is:

[183 1 0]

The numbers one and zero are numerical equivalents of Boolean "True" and "False".

Numpy arrays cannot be changed once created. Although, new arrays can be made by manipulating the existing array.

err = np.append(arr, 3347)

print(err)

The output is:

[36   69   3347]

Reminder: leading zeros are not allowed in Python numbers anymore.

frr = np.array([14.7, 3.00])

print(frr)

Which gives us the output of:

[14.7 3.]

Appending data in a multidimensional array creates a new one-dimensional array.

grr = np.append(brr[0], 4) #new one dimensional array

print(grr)

The output is:

[75 53 95 51 4]

Removing an element from an array also leads to creation of a new array.

hrr = np.delete(grr, 3) #this method removes an element by finding it through its index. 3 has index of 4 in grr

print(hrr)

The output is:

[75 53 95 51]

There are many standard arrays that are used in various applications. The numpy library has special methods to create such arrays quickly.

zrr = np.zeros((2, 3)) #this will create an array with all elements zero in floating type

print(zrr)

zrrint = np.zeros((2, 3), dtype=int) # if we want to create an array of integer zeros

print(zrrint)

The outputs are.

[[0. 0. 0.]

 [0. 0. 0.]]

[[0 0 0]

 [0 0 0]]

We can also create a three-dimensional (3D) array.

threDrr = np.ones((2, 3, 3)) #this is taken as a multidimensional (3D) array consisting of two 2D arrays each of 3 rows and 3 columns

print(threDrr)

The output is:

[[[1. 1. 1.]

 [1. 1. 1.]

 [1. 1. 1.]]

[[1. 1. 1.]

[1. 1. 1.]

[1. 1. 1.]]]

If you want the elements of an array to follow a specific sequence, we can do that with a couple of methods.

seqrr = np.arange(0, 101, 10) #create an array with elements in a sequence by setting how much interval should be added between each element

print(seqrr)

linrr = np.linspace(0, 100, 11, dtype=int) #create an array with elements in a sequence by setting how many elements are needed in the array. Python automatically sets the appropriate step value

print(linArr)

The outputs is:

[0 10 20 30 40 50 60 70 80 90 100]

[0 10 20 30 40 50 60 70 80 90 100]

A few things to remember:

1. linspace() includes the maximum value in the sequence unlike the arange() which works like the range()
2. linspace() creates sequence in floating type numbers unless instructed otherwise just like in our example ("dtype=int").

To create an array with all element positions taken by the same number, we use the full() method.

fillrr = np.full((2,2), 7)

print(fillrr)

The code outputs the following:

[[7 7]

[7 7]]

As stated earlier, the numpy arrays are the closest implementation of matrices in Python. As such, there are various methods available in numpy library to perform matrix operations. Identity matrix are used frequently in matrix operations and we can create an identity matrix as a numpy array with a single line of code.

idrr = np.eye(2, dtype=int) #identity matrix always has the same number of rows and columns (i.e they are square matrices) and one of the diagonals has all elements set as '1'

print(idrr)

The output is:

```
[[1 0]

 [0 1]]
```

The numpy library has randomizer methods that generate random integer or float numbers to fill in an array. This is particularly useful when you want to create several arrays to test a script.

randIrr = np.random.randint(1, 15, size=(4, 4)) #1 and 15 are minimum and maximum limits so numpy will generate random numbers that lie within this range. (4, 4) will be the size of new array created

print(randIrr)

randFrr = np.random.random((3, 3)) #this time we didn't give any limits and the returned array will have a size of 3 rows and 3 columns

print(randFrr)

The output of the above code is below. You will see different output when you run the code because randomizer will output different numbers.

```
[[ 7 8 9 9]
```

[5 14 7 2]

[3 2 9 13]

[1 3 8 8]]

[[0.29457456 0.83468151 0.35160973]

[0.29059067 0.896374607 0.38657756]

[0.5831767 0.54060431 0.94441115]]

Below are some examples of operations you can perform on numpy arrays.

arr = np.array([

    [7, 8, 9],

    [9, 8, 7]

])

brr = np.array([

    [1, 2, 3],

    [3, 2, 1]

])

#addition

print(np.add(arr, brr)) # this is a scalar addition which we can also do with '+' operator like arr + brr

#subtraction

print(np.subtract(arr, brr)) #again a scalar operation hence aArr - bArr also gives same output

#multiplication

print(np.multiply(arr, brr)) # element by element (scalar) which is also arr * brr

#division

print(np.divide(arr, brr)) #also aArr / bArr because a scalar operation

The output of the above script is:

[[8 10 12]

 [12 10 8]]

[[6 6 6]

[6 6 6]]

[[7 16 27]

 [27 16 7]]

[[7. 4. 3.]

 [3. 4. 7.]]

We can also perform dot and cross products on arrays.

drr = np.array([

    [7, 8, 9],

    [9, 8, 7],

    [4, 5, 6]

])

err = np.array([

    [1, 2, 3],

    [3, 2, 1],

    [4, 5, 6]

])

#dot product

print(np.dot(drr, err))

#cross product

print(np.cross(drr, err))

The outputs are:

[[67 75 83]

 [61 69 77]

 [43 48 53]]

[[6 -12 6]

 [-6 12 -6]

 [0 0 0]]

Dot and cross products play a significant role in vector theory. You can read about it online. To transpose an array, we have the property 'T'.

print(drr.T)

The output is:

[[7 9 4]

 [8 8 5]

[9 7 6]]

Finding different kind of sums results in many interesting values.

print(np.sum(drr)) #sum all elements in an array, returns a single value

#sum the elements in either the rows or columns

print(np.sum(drr, axis=0)) # axis = 0 means we sum all elements in the rows

print(np.sum(drr, axis=1)) # axis = 1 means we sum all elements in the columns

Here are the outputs:

63

[20 21 22]

[24 24 15]

Chapter 5: Advanced Data Analysis

Chapter 4 has a lot of information on data analysis, but Python supports much more advanced data analysis tools.

Using pandas Framework

When it comes to data analysis, Python has another ace up its sleeve: the "pandas" framework. The framework provides two new data structures, "series" and "dataframe". The framework also provides out of the box data acquisition techniques for all popular data sources such as spreadsheets and CSV files. Without wasting another minute, let us dive into the concepts and working of this library.

pandas Series

The "pandas" Series data structure has features combined from both dictionaries and numpy arrays. It is an advanced way of declaring and manipulating a group of numbers with a specific sequence.

from pandas import Series

import pandas as pd

```
asr = Series([1, 7, 9]) # the syntax closely resembles that of a
numpy array
```

```
print(asr)
```

```
print(asr.values) # get all values as a list just like a dictionary
```

```
print(asr.index) # get all indices as a list, we can say they are the
```
"keys" of this data structure - another resemblance with a dictionary. Since we didn't set specific keys, the values were assigned number indices

```
kes = ['1st', '2nd', '3rd']
```

```
bsr = Series([3, 9, 7], index=kes) # we can set custom "keys" for
the data structure
```

```
print(bsr)
```

```
print(bsr[1st']) # get element using key just like a dictionary
```

```
print(bsr[0]) # get element using index just like any other data
structure in Python
```

```python
adic = {

    '1st': 245,

    '2nd': 265,

    '3rd': 888

}

csr = Series(adic)  # creating a series from a dictionary is very easy


print(csr)


ind = ['1st', '2nd', '4th']

dsr = Series(adic, index=ind)  # if a key isn't in the original dictionary (4th isn't present in this case), "NaN" is added as value for that key


print(dsr)


print(pd.isnull(dsr['4th']))  # check if an element has "NaN" value, in this case the output will be Boolean True
```

dsr.name = 'Info' # give name to entire series

dSer.index.name = 'Key' # give name to the index/key column

print(dsr)

dsr.index = ['un', 'deux', 'nul'] # we can change the keys of a series, this is a major upgrade from a dictionary where the keys are immutable. Changing the indices/keys removes any name we have given to the index column

print(dsr)

The outputs of the above "print" statements are.

```
0       1

1       7

2       9

dtype: int64

[1 7 9]

RangeIndex(start=0, stop=3, step=1)
```

1st   3

2nd   9

3rd   7

dtype: int64

3

3

1st   245

2nd   265

3rd   888

dtype: int64

1st   245.0

2nd   265.0

4th   NaN

dtype: float64

True

Key

1st   245.0

2nd   265.0

4th   NaN

Name: Info, dtype: float64

un   245.0

deux   265.0

nul   NaN

Name: Info, dtype: float64

<u>Time series</u>

A quantity that can be measured with respect to time can be represented as a time series - a special type of pandas Series where the index column has time values. If you take a look at any log file, you will notice how each entry is preceded by a timestamp showing when that particular log entry was made. The logs are sorted according to time, either latest to earliest or vice versa.

The pandas framework provides a lot of methods to manipulate time series that are beneficial in various applications. All the rules of pandas series work on time series but we need to understand how Python works with dates before we can move on to the next topic. The below script shows different manipulation techniques of time.

```python
from datetime import datetime

from datetime import timedelta

from dateutil.parser import parse


nowTime = datetime.now()  # get the current date and time value


print(now)  # the returned data has the current date and time
including the year, month, day, hour, minutes, seconds,
milliseconds, and microseconds values


nxT = nowTime + timedelta(12)   # timedelta() creates time
object that can be used to jump time according to the arguments
provided. In this case, we want to get the date and time 12 days
from current date and time


print(nxT)


print(str(nxT))  # string looks the same as datatime format
```

```python
print(nxT.strftime('%Y-%m-%d'))  # reformat datetime value


impTime = '2016-11-09'


print(datetime.strptime(impTime, '%Y-%m-%d'))   # format string as datetime


print(parse(impTime))  # when the string has a format that's the same as a datetime format, we don't need to use strptime() and declare a format. We can use parse() to quickly convert such string to a datetime object


print(parse(ranTime, dayfirst=True))  # the dayfirst argument swaps the position of months with days to follow the international date and time convention.
```

The respective outputs of the above print statements are:

2019-10-30 08:10:25.841218

2019-11-11 08:10:25.841218

2019-11-11 08:10:25.841218

2019-11-11

2016-11-09 00:00:00

2016-11-09 00:00:00

2016-09-11 00:00:00

pandas DataFrames

The dataframe is the most advanced data structure in Python and can handle gigabytes of data processing without slowing down the computer. Thanks to the process optimizations done behind the scenes, pandas dataframes are ideal for all data analysis and machine learning applications.

Let us take a look at some basic dataframe manipulation.

```
import pandas as pd


monLst = ['Jan', 'Feb', 'Mar', 'Apr']


quartrStats = {

    "Ad Spending": [12000, 24000, 9000, 11000],

    "Revenue": [55000, 15745, 21000, 47314],
```

"Store Manager": ['Jones', 'Joshua', 'Marcus', 'Smith']

}

dfnew = pd.DataFrame(quartrStats, index=monLst) # assigning custom keys/indices to the dataframe

dfalt = pd.DataFrame(quartrStats) # creating a dataframe with default keys/indices

dfalt.set_index('Store Manager', inplace=True) # assign an existing data column as the index column, "inplace" argument makes sure the original dataframe is overwritten with the change. By default, whenever a change is made to a dataframe, the original is kept intact and a new dataframe is created to carry the changes.

print(dfnew)

print(dfalt)

The output is:

	Ad Spending	Revenue	Store Manager
Jan	12000	55000	Jones
Feb	24000	15745	Joshua

Mar	9000	21000	Marcus
Apr	11000	47314	Smith

	Ad Spending	Revenue
Store Manager		
Jones	12000	55000
Joshua	24000	15745
Marcus	9000	21000
Smith	11000	47314

We can extract a single column or a set of columns from a dataframe. Converting a dataframe column to a numpy array is also straightforward, which leads to the million dollar question: can we use a numpy array to create a dataframe? The code below actually performs all these tasks.

print(dfnew['Revenue']) # a new series is created with the same index as the source dataframe

print(dfnew.Revenue) # this will only work if the column header has no spaces, therefore it is a good idea to use underscores '_' or dashes '-' instead of spaces when adding column header in a dataframe

print(dfnew[['Ad Spending', 'Revenue']])   # use multiple columns of a dataframe to create a new dataframe with the same index as the original one

print(np.array(dfnew[['Ad Spending', 'Revenue']]))   # convert the newly extracted dataframe to a numpy array

The output of the above script is:

Jan   55000

Feb   15745

Mar   21000

Apr   47314

Name: Revenue, dtype: int64

Jan   55000

Feb   15745

Mar   21000

Apr   47314

Name: Revenue, dtype: int64

    Ad Spending Revenue

```
Jan  12000           55000

Feb   24000          15745

Mar  9000            21000

Apr  11000            47314

[[12000        55000]

 [24000        15745]

 [9000         21000]

 [11000        47314]]
```

Parsing Data

"pandas" framework supports easy and fast data retrieval from various resources without the need of another library. Search the Internet for "AAPL", which is the stock price data of the global tech giant Apple Inc.. You can use a data aggregator service like Quandl. Let's download the data to the Python folder in the "csv" format. The following code will read all the downloaded data.

```
import pandas as pd


# read local CSV data
```

```python
datF = pd.read_csv('EOD-AAPL.csv') # this is the downloaded
CSV file, replace it with the name of your file


datF.set_index('Date', inplace=True) # making date column the
index of the dataframe

datF.to_csv("newF.csv")  # create a new CSV file and transfer all
the dataframe data


print(datF.head()) # print the first five rows of the dataframe


datF = pd.read_csv('newF.csv', index_col=0)  # this is how to set
a specific data column as an index column during data read using
the column index. Zero index means the first column


print(datF.head())


datF.rename(columns={'Open': 'Open_Price'}, inplace=True) #
we can rename a single or multiple columns by passing the
values as key-value pairs of a dictionary to the rename() method
```

```
print(datF.head())
```

```
datF.to_csv("newFn.csv", header=False)  # save data to a CSV
file while ignoring the column header
```

```
datF = pd.read_csv('newFn.csv', names=['Date', 'Open', 'High',
'Low', 'Close', 'Volume', 'Dividend', 'Split', 'Adj_Open',
'Adj_High', 'Adj_Low', 'Adj_Close', 'Adj_Volume'],
index_col=0)  # we can also set the column names when reading
data from a CSV file
```

```
print(datF.head())
```

The output of the above script is:

```
        Open       High  Low ...      Adj_Low    Adj_Close
Adj_Volume
Date                          ...

2017-12-28 171.00 171.850 170.480 ... 165.957609 166.541693
16480187.0

2017-12-27 170.10 170.780 169.710 ... 165.208036 166.074426
21498213.0
```

2017-12-26 170.80 171.470 169.679 ... 165.177858 166.045222
33185536.0

2017-12-22   174.68   175.424   174.500   ...   169.870969
170.367440 16349444.0

2017-12-21 174.17 176.020 174.100 ... 169.481580 170.367440
20949896.0

[5 rows x 12 columns]

       Open High Low ...   Adj_Low   Adj_Close
Adj_Volume

Date   ...

2017-12-28 171.00 171.850 170.480 ... 165.957609 166.541693
16480187.0

2017-12-27 170.10 170.780 169.710 ... 165.208036 166.074426
21498213.0

2017-12-26 170.80 171.470 169.679 ... 165.177858 166.045222
33185536.0

2017-12-22   174.68   175.424   174.500   ...   169.870969
170.367440 16349444.0

2017-12-21 174.17 176.020 174.100 ... 169.481580 170.367440 20949896.0

[5 rows x 12 columns]

	Open_Price	High	Low	...	Adj_Low	Adj_Close	Adj	Volume
Date				...				
2017-12-28	171.00	171.850	170.480	...	165.957609	166.541693	16480187.0	
2017-12-27	170.10	170.780	169.710	...	165.208036	166.074426	21498213.0	
2017-12-26	170.80	171.470	169.679	...	165.177858	166.045222	33185536.0	
2017-12-22	174.68	175.424	174.500	...	169.870969	170.367440	16349444.0	
2017-12-21	174.17	176.020	174.100	...	169.481580	170.367440	20949896.0	

[5 rows x 12 columns]

```
       Open High  Low  ...        Adj_Low              Adj_Close
Adj_Volume

Date                    ...

2017-12-28 171.00 171.850 170.480 ... 165.957609 166.541693
16480187.0

2017-12-27 170.10 170.780 169.710 ... 165.208036 166.074426
21498213.0

2017-12-26 170.80 171.470 169.679 ... 165.177858 166.045222
33185536.0

2017-12-22   174.68   175.424   174.500   ...   169.870969
170.367440 16349444.0

2017-12-21 174.17 176.020 174.100 ... 169.481580 170.367440
20949896.0


[5 rows x 12 columns]
```

Write data

We can use pandas framework to write data in various common formats. Let's look at an example where we write data to an HTML file.

datF.to_html('datF.html') # such as simple implementation

You can view this new HTML file by opening it up in any browser. To view the markup of this page, you can open it using any text editor. You can also use the inspection tool of the browser to view the markup.

Internet Scraping using pandas

We can scrape websites using the pandas special library "pandas_datareader". We need to install it using pip before we can use it.

pip install pandas_datareader

Run the following script after the installation has completed.

import pandas_datareader.data as web

rngstrt = '1/1/2009'

rngstp = '5/5/2015'

dfnew = web.get_data_yahoo('GOOG', rngstrt, rngstp)

print(dfnew.head()) #output first five rows of the dataframe

The output will be:

	High	Low	...	Volume	Adj Close
Date			...		
2009-01-02	160.309128	152.179596	...	7248000.0	160.060059
2009-01-05	165.001541	156.911850	...	9814500.0	163.412491
2009-01-06	169.763687	162.585587	...	12898500.0	166.406265
2009-01-07	164.837143	158.779861	...	9022600.0	160.403763
2009-01-08	161.987823	158.077484	...	7228300.0	161.987823

[5 rows x 6 columns]

Visualize Data

It is very difficult to even read thousands of lines of data, let alone find a relationship in the data. But, this becomes easier when we plot charts using the data. Many trends and system behaviours becomes instantly recognizable.

Plot a Chart

There are several libraries to create graphs using Python. We have already seen matplotlib in action. Let's work with pygal now. We can install this library using pip.

pip install pygal

All the codes in this section are inspired from the Pygal official documentation available at http://www.pygal.org/en/stable/documentation/. Here's a simple line chart generating code.

import pygal

from pygal.style import LightenStyle

dark_lighten_style = LightenStyle('#000000')

bar_chart = pygal.Bar(style=dark_lighten_style)

bar_chart.add('Fibonacci', [0, 1, 1, 2, 3, 5, 8, 13, 21, 34, 55])

bar_chart.add('Padovan', [1, 1, 1, 2, 2, 3, 4, 5, 7, 9, 12])

bar_chart.render() # show the chart

bar_chart.render_to_file('bar_chart.svg') # save the chart as svg image file

Here is the output chart:

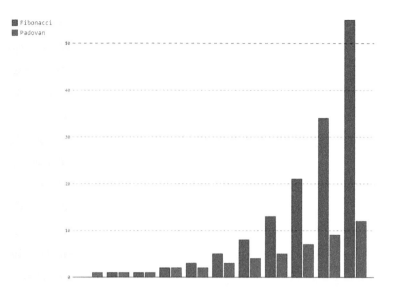

We are using the LightenStyle() to create a grayscale color scheme with the base color of black (hex code: #000000). We pass that custom color scheme as the style to be used for the chart. Pygal smartly chooses the best grayscale shades to plot the chart.

Let's create a pie chart.

import pygal

from pygal.style import LightenStyle

dark_lighten_style = LightenStyle('#000000')

pie_chart = pygal.Pie(style=dark_lighten_style)

```
pie_chart.title = 'Browser usage in February 2012 (in %)'

pie_chart.add('IE', 19.5)

pie_chart.add('Firefox', 36.6)

pie_chart.add('Chrome', 36.3)

pie_chart.add('Safari', 4.5)

pie_chart.add('Opera', 2.3)

pie_chart.render() # show the pie chart

pie_chart.render_to_file('pie_chart.svg') # save the chart as svg
image file
```

The output chart looks like the following:

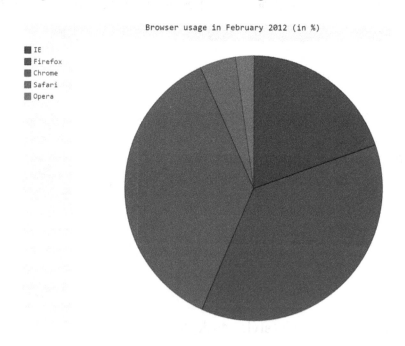

Notice how the Pygal is taking care of styling the chart legend and title. If we had used "matplotlib", we would have had to set all these things ourselves.

There are so many options available with Pygal and you should explore the library further. One last example we are going to cover is to create a box chart, which is a very common sight in financial applications, especially trading. Here's the complete code.

```
import pygal

from pygal.style import LightenStyle


dark_lighten_style = LightenStyle('#000000')


box_plot = pygal.Box(style=dark_lighten_style)

box_plot.title = 'V8 benchmark results'

box_plot.add('Chrome', [6395, 8212, 7520, 7218, 12464, 1660, 2123, 8607])

box_plot.add('Firefox', [7473, 8099, 11700, 2651, 6361, 1044, 3797, 9450])
```

box_plot.add('Opera', [3472, 2933, 4203, 5229, 5810, 1828, 9013, 4669])

box_plot.add('IE', [43, 41, 59, 79, 144, 136, 34, 102])

box_plot.render_to_file('box_chart.svg') # save the chart as svg image file

The output:

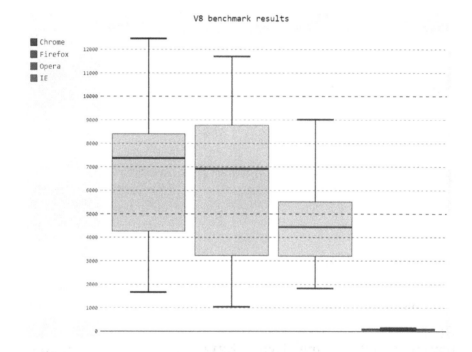

Final note: "svg" format is still a relatively new format and you might not be able to open images with this format on your Windows computer. You can use your Google Chrome browser to view "svg" images.

Chapter 6: Web Applications with Python

Django is an external library widely used to create web applications using Python. It's used by online giants like Instagram and Pinterest. The codes in this chapter are inspired by the work of YouTube channel "techwithtim".

Setup Django

Let's install using pip.

pip install django

Now, let's create our first application. We need to setup the initiation files as a project that will serve as the environment of the application. Using the command prompt, navigate to a folder that will house the application files. To setup the application project, run the following command.

django-admin startproject firstapp

"firstapp" is the name of our first project. To see if the project is setup correctly, run the following command on the command prompt. Make sure you have changed current directory to the correct folder.

python manage.py runserver

To create an application inside this project environment, run the following command.

python manage.py startapp firstapp

Note that the project name and application can be different. To change the different application pages (called views), move to the application folder and change the file named "views.py".

views.py file

from django.http import HttpResponse

def index(request):

   return HttpResponse("This is a brand new web app!")

We need to create a URL to access this view. Let's create a new Python file "urls.py" in the application directory. Add the following code to it:

urls.py

from django.urls import path

from . import views

```
urlpatterns = [

    path('', views.index, name='index'),

]
```

There is another "urls.py" if you navigate inside the application directory. Modify it so it has the following code.

```
# urls.py

from django.contrib import admin

from django.urls import include, path


urlpatterns = [

    path('', include(firstapp.urls')), # firstapp is the name of our app

    path('admin/', admin.site.urls),

]
```

You can view this web application by going to the following URL in the browser.

127.0.0.1:8000

First Web Application

Every web application that deals with data usually has a database to store information. We are going to set up an SQLite3 database for our application.

Go to the "settings.py" file inside the web application folder and add one line in the section.

Application definition

INSTALLED_APPS = [

   'django.contrib.admin',

   'django.contrib.auth',

   'django.contrib.contenttypes',

   'django.contrib.sessions',

   'django.contrib.messages',

   'django.contrib.staticfiles',

  'firstapp.apps.FirstappConfig', # <- add this line, firstapp is the name of our application

]

Now using command prompt, let's move to the directory containing "manage.py" and run the following command.

python manage.py migrate

This command will create an empty database in the application directory. We can add information in the database when needed.

Let's create an admin page. We will first need to setup the database correctly. Run the following command but first make sure you are in the directory that contains "manage.py"

python manage.py createsuperuser

You will now see an admin dashboard created. To view this newly created admin page, go to the following link

127.0.0.1:8000/admin

The admin dashboard is usually created to view the details of the web application. Let's put the database details in the dashboard by adding following block of code in the "admin.py".

from django.contrib import admin

from .models import ToDoList, Item

Register your models here.

admin.site.register(ToDoList)

admin.site.register(Item)

Let's get a little fancy and start with HTML templates. Templates are a well-known concept in web designing nowadays. Templates are HTML files that can be shared by different pages that should have the same design and content. Of course, you can overwrite specific details of a template to show different stuff on a specific page. For this project, we are going to create a base HTML template file and then create two more HTML files that share that template. In the main folder, create a base.html and add the following code to it.

```
<html>

<head>

    <title>{% block title %}Tim's Site{% endblock %}</title>

</head>


<body>

    <div id="content" name="content">

                    {% block content %}

                    {% endblock %}

    </div>
```

```
</body>

</html>
```

Notice the places where we have used placeholders wrapped in "{% %}". We must include the closing "{% endblock %}" placeholder. We will insert custom content here when we create specific HTML files. Let's create home.html with the following code. The first line tells Python what parent template to use for this HTML file.

```
{% extends 'main/base.html' %}


{% block title %}

Home

{% endblock %}


{% block content %}

<h1>Home Page</h1>

{% endblock %}
```

Let's create the other HTML file that will use some Python conditional statements to generate the code. Variable names are enclosed in double spaces "{{}}".

```
{% extends "main/base.html" %}

{% block title %}View List{% endblock %}


{% block content %}
                    <h1>{{ls.name}}</h1>
                    <ul>
                    {% for item in ls.item_set.all %}
                        {% if item.complete == True %}
                            <li>{{item.text}} -
COMPLETE</li>
                        {% else %}
                            <li>{{item.text}} -
INCOMPLETE</li>
                        {% endif %}
                    {% endfor %}
                    </ul>
{% endblock %}
```

We have to update the "views.py" and "url.py" to start showing the new HTML files (pages) on the web application.

views.py

```python
from django.shortcuts import render

from django.http import HttpResponse

from .models import ToDoList, Item

# Create your views here.


def index(response, id):

    ls = ToDoList.objects.get(id=id)

    return render(response,
"main/list.html", {"ls":ls})


def home(response):

    return render(response,
"main/home.html", {})
```

The empty {} is used to pass a dictionary containing the variables to the templates.

urls.py

```python
from django.urls import path

from . import views

urlpatterns = [

path("<int:id>", views.index, name="index"),

path("", views.home, name="home")

]
```

After adding simple HTML page, let's add a simple HTML form. Let's create another template "formCreate.html" with the following code.

```
{% extends 'main/base.html' %}

{% block title %}

Create New List

{% endblock %}


{% block content %}
```

<h3>Create a New To Do List</h3>

```
<br>

<form method="post"
action="/create/" class="form-group">

{% csrf_token %}

<div class="input-group mb-3">

<div class="input-
group-prepend">

<button name="save"
type="submit" class="btn btn-success">Create</button>

</div>

{{form.name}}

</div>

</form>

{% endblock %}
```

We must use "{% csrf_token %}" in every form we create because it adds necessary security measures to the form. The {{form}} is set when creating a form using this template, you can say it will be the ID of that form.

Let's update "views.py" for this form.

```python
def create(response):

    return render(response, "main/formCreate.html", {"form":
form}) # goes inside views.py
```

We also have to update the "urls.py".

```python
from django.urls import path

from . import views


urlpatterns = [

path("", views.home, name="home"),

path("home/", views.home, name="home"),

path("create/", views.create, name="index"),

path("<int:id>", views.index, name="index"),

]
```

We have to add the following code block to "views.py" to create a form instance using the formCreate.html

```python
...

from .forms import CreateNewList

...
```

```python
def create(request):

    form = CreateNewList()

    return render(request, "main/formCreate.html", {"form":
form})
```

Now, we have built the form but there's no way to retrieve information from the form. Let's understand some HTML concepts.

POST Request: When information needs to be passed from the front-end (website interface) to the backend (server/database), POST HTML request is used. This is a secured and encrypted request.

GET Request: When information needs to be passed from the backend (server/database) to the front-end (website interface), POST HTML request is used. This is an open and unsecured request.

We have to add the POST and GET requests to the "views.py".

```python
def create(response):

    if response.method == "POST":

        form = CreateNewList(response.POST)
```

```python
if form.is_valid():

    n = form.cleaned_data["name"]

    t = ToDoList(name=n)

    t.save()


    return HttpResponseRedirect("/%i" %t.id)


else:

    form = CreateNewList()


return render(response, "main/formCreate.html", {"form":form})
```

We can add Bootstrap stylizing and JavaScript customization to the django web application. What's Bootstrap? It's a style library available for free to make web applications responsive. To add these features, we have the change the base.html file. Replace <home> with <!doctype html> and add the following lines between <head> and </head>.

```html
<meta charset="utf-8">
```

```html
<meta name="viewport" content="width=device-width, initial-scale=1, shrink-to-fit=no">
```

```html
<link rel="stylesheet" href="https://stackpath.bootstrapcdn.com/bootstrap/4.3.1/css/bootstrap.min.css">
```

Let's add the following codes just before the </body> tag.

```html
<script src="https://code.jquery.com/jquery-3.3.1.slim.min.js"></script>
```

```html
<script src="https://cdnjs.cloudflare.com/ajax/libs/popper.js/1.14.7/umd/popper.min.js"></script>
```

```html
<script src="https://stackpath.bootstrapcdn.com/bootstrap/4.3.1/js/bootstrap.min.js"></script>
```

Now, we can add Bootstrap classes to the elements to make them better looking without writing our own stylesheet. We can also write JavaScript and jQuery codes for added dynamics.

The possibilities are endless with Django Python framework and there are even more Django extension libraries available. For example, to add the capability of user sign up we can use "django-crispy-forms" external library.

Chapter 7: GUI and Computer Peripheral Control

Working with command prompt might look nerdy cool at first, but it's not a good way to take inputs from users in today's world. Everything requires a visual touch. Python provides considerable resources and tools to create a Graphical User Interface (GUI). Let's create some fun things.

Create GUI for File and Folder Selection

We can give users the ability to visually navigate to a file and folder for various file operations. This is enabled by using Python's standard GUI library, Tkinter, a unique name with powerful GUI capabilities. The code enables a user to select a file so we can extract the file path.

import Tkinter

import tkFileDialog

rut = Tkinter.Tk()

rut.withdraw() #hide the tkinter window

```
fileAcc                                              =
tkFileDialog.askopenfile(parent=root,initialdir="/",mode='rb',t
itle='Pick any file')

if fileAcc != None:

            print fileAcc.name
```

We can restrict the file type the user can select through the GUI. Let's pass more arguments to the askopenfile() method.

```
fileAcc                                              =
tkFileDialog.askopenfile(parent=root,initialdir="/",mode='rb',t
itle='Pick any file',filetypes = (("All Excel files","*.xl*"),("All
files","*.*")))
```

Taking Control of Keyboard and Mouse

We are entering the domain of programming that can be attributed to hacking. Taking control of a keyboard and mouse can be used for nefarious purposes. But, there are benefits to teaching this topic.

1. Clients want to test their web applications. You can write a Python program to automatically test their web app in several ways. One script can do the work of an entire QA team.

2. You want to automate some tasks on your computer that

will take too much time to directly integrate with Python. For example, instead of using an Excel related library to open an Excel spreadsheet, you can use this technique to achieve the same results.

The external library "pyautogui" helps us in this respect. No other library is required for a Windows system.

pip install pyautogui

When "pyautogui" takes control of either keyboard or mouse, it blocks your access to those devices. If you make an error in the script, you might lose access to the input devices unless you hard reset the computer. Don't worry, "pyautogui" provides pause and failsafe features. "Failsafe" is triggered when the mouse hits the top right corner of the screen and device control is returned to the user. The "pause" creates a delay so you can stop the program execution if need be. Note that once you lose control of your keyboard, you won't be able to use the CTRL+C to break the program execution.

import pyautogui

pyautogui.FAILSAFE = True

pyautogui.PAUSE = 1

Before we automate the keyboard or mouse, we need to know what is the size of the screen so we would know the limits.

```
pyautogui.size()
```

Mouse automation

```
for xy in range(8):
    pyautogui.moveTo(50, 50, duration=0.5)
    pyautogui.moveTo(100, 50, duration=0.5)
    pyautogui.moveTo(100, 100, duration=0.5)
    pyautogui.moveTo(50, 100, duration=0.5)
```

The above code moves the mouse cursor in a square shape pattern in the clockwise direction. Each movement will occur in 0.25 seconds so it will be visible for you to see the movement. In the above example, we have used fixed coordinates. We can use relative coordinates also, just like we have used below.

```
pyautogui.moveRel(50, 50, duration=0.5)
```

All this wouldn't make sense if we don't know what the current position of the cursor is.

```
pyautogui.position()
```

We can also automate mouse click like this.

pyautogui.click(7, 3)

We have to provide fixed coordinates to the click() method, not relative. The click() method mimics the left mouse button being pressed and released. We can use "pyautogui.mouseDown()" and "pyautogui,mouseUp()" to automate either pressing or releasing of the mouse button. The "pyautogui.doubleClick()" simulates a double click. You can use "pyautogui.rightClick()" to simulate a right click and a middle button click with "pyautogui.middleClick()".

There are two methods "pyautogui,dragTo()" and "pyautogui.dragRel()" to automate dragging, both of which work in a similar manner as moveTo() and moveRel() methods we have seen before. The difference is that with drag modules, the mouse button is also left clicked to force a drag instead of just moving the cursor.

Newer mice come with the scrolling wheel and it can be automated too with the scroll() method. You have to pass an integer value that pyatuogui uses to determine how much the screen has to move up or down. The integer values will result in different scroll movement on different screen sizes and computer systems. One important thing to remember: the scroll() method is relative, which means the scrolling will start from the current position of the mouse.

We can also take a screenshot with "pyautogui" library using the screenshot() method.

viewfreeze = pyautogui.screenshot()

Keyboard automation

Keyboard actions can be simulated using the rhe "pyautogui.typewrite()" method. Depending upon the application currently active, the same key presses might result in different actions. Here is an example of typing a string with each keystroke happening after a delay.

pyautogui.typewrite("Dora The Explorer!", 0.71)

We can also automate keyboard actions such as adding a new line by passing a special code in the typewrite(). Here's a table of codes assigned to different keyboard actions.

Key string	Meaning
'a', 'b', 'c', 'A', 'B', 'C', '1', '2', '3', '!', '@', '#', and so on	keys for single characters
'enter' (or 'return'or '\n')	ENTER key
'esc'	ESC key

'shiftleft', 'shiftright'	Left and Right SHIFT keys
'altleft','altright'	Left and Right ALT keys
'ctrlleft','ctrlright'	Left and Right CTRL keys
'tab' or '\t'	TAB key
'backspace','delete'	BACKSPACE and DELETE keys
'pageup','pagedown'	PAGE UP and PAGE DOWN keys
'home','end'	HOME and END keys
'up','down','left','right'	ARROW keys
'f1','f2','f3' and so on...	F1 to F12 keys
'volumeup','volumedown','mute'	VOLUME control keys. Your keyboard might have these keys but your system might still be able to recognize these commands

'pause'	PAUSE key
'capslock','numlock','scrolllock'	LOCK keys
'insert'	INSERT or INS key
'printscreen'	PRINT SCREEN or PRTSC key
'winleft','winright'	Left and Right WIN keys (Windows only)

We can combine different keypresses to type a special character. The following lines of code result in '%' getting typed.

pyautogui.keyDown('shift')

pyautogui.press('5')

pyautogui.keyUp('shift')

The hotkey() method is very useful because we can perform system operations like copy and paste by mimicking the simultaneous press of CTRL and another key. For example, here is how we can automate a copy command.

pyautogui.hotkey('ctrl','c')

Chapter 8: Gaming with Python

Gaming is not new, it's as old as the computers. It's one of the biggest industries related to computers and programming in terms of revenue and salaries. With the advent of smartphones, more and more independent developers are launching games in the hope of hitting the jackpot. The codes in the chapter are inspired by the work of Youtube channel "techwithtim".

Python provides a way to create 2D games and we are going to take a look at it now.

PyGame Introduction

The "pygame" is an excellent external library for building games. Let's install "pygame".

pip install pygame

The following code shows the basics of setting up a game environment. Comments are added to explain important points.

import pygame

pygame.init() # start the gaming engine

win = pygame.display.set_mode((500, 500)) # This line creates a window of 500 width, 500 height

```python
pygame.display.set_caption("First Game") # change window
name
```

Initializing the character attributes.

```python
x = 50

y = 50

width = 40

height = 60

vel = 5

# setting an infinite loop

run = True


while run:

    pygame.time.delay(100)


    for event in pygame.event.get():

        if event.type == pygame.QUIT:

            run = False # end game loop only in this condition
```

```python
keys = pygame.key.get_pressed() # check if a keyboard key is
pressed and which key it is

# move the character wrt to the keyboard key pressed

if keys[pygame.K_LEFT]:

    x -= vel


if keys[pygame.K_RIGHT]:

    x += vel


if keys[pygame.K_UP]:

    y -= vel


if keys[pygame.K_DOWN]:

    y += vel


win.fill((0,0,0))  # Fills the screen with black so there's no tail
of the character
```

```
pygame.draw.rect(win, (255,0,0), (x, y, width, height)) # draw
the character
```

```
pygame.display.update() # update the display on every loop
execution
```

```
pygame.quit() # stop the game when loop is broken
```

Creating a 2D Game

We are going to create the popular Tetris game that will become
more difficult over time.

Part #1: Initialization

We have to create the game environment before we can start
coding the mechanics and other details. Here's the complete
code that initializes the Tetris game environment. Comments are
added where necessary to describe certain aspects.

```
import pygame

import random


# creating the data structure for pieces
```

```python
# setting up global vars

# functions

# - create_grid

# - draw_grid

# - draw_window

# - rotating shape in main

# - setting up the main



"""

10 x 20 square grid

shapes: S, Z, I, O, J, L, T

represented in order by 0 - 6

"""



pygame.font.init()



# GLOBALS VARS
```

```
s_width = 800

s_height = 700

play_width = 300  # meaning 300 // 10 = 30 width per block

play_height = 600  # meaning 600 // 20 = 20 height per block

block_size = 30


top_left_x = (s_width - play_width) // 2

top_left_y = s_height - play_height
```

SHAPE FORMATS


```
S = [['.....',
      '.....',
      '..00..',
      '.00...',
      '.....'],
```

```
['.....',

'..o..',

'..oo.',

'...o.',

'.....']]


Z = [['.....',

'.....',

'.oo..',

'..oo.',

'.....'],

['.....',

'..o..',

'.oo..',

'.o...',

'.....']]
```

```
I = [['..o..',

      '..o..',

      '..o..',

      '..o..',

      '.....'],

     ['.....',

      'oooo.',

      '.....',

      '.....',

      '.....']]


O = [['.....',

      '.....',

      '.oo..',

      '.oo..',

      '.....']]
```

```
J = [[['.....',

'.o...',

'.ooo.',

'.....',

'.....'],

['.....',

'..oo.',

'..o..',

'..o..',

'.....'],

['.....',

'.....',

'.ooo.',

'...o.',

'.....'],

['.....',

'..o..',
```

```
    '..o..',

    '.oo..',

    '.....']]


L = [['.....',

    '...o.',

    '.ooo.',

    '.....',

    '.....'],

   ['.....',

    '..o..',

    '..o..',

    '..oo.',

    '.....'],

   ['.....',

    '.....',

    '.ooo.',
```

'.o...',

'.....'],

['.....',

'.oo..',

'..o..',

'..o..',

'.....']]

T = [['.....',

'..o..',

'.ooo.',

'.....',

'.....'],

['.....',

'..o..',

'..oo.',

'..o..',

```
      '.....'],

    ['.....',

     '.....',

     '.ooo.',

     '..o..',

     '.....'],

    ['.....',

     '..o..',

     '.oo..',

     '..o..',

     '.....']]

shapes = [S, Z, I, O, J, L, T]

shape_colors = [(0, 255, 0), (255, 0, 0), (0, 255, 255), (255, 255,
0), (255, 165, 0), (0, 0, 255), (128, 0, 128)]

# index 0 - 6 represent shape
```

```python
class Piece(object):

    pass


def create_grid(locked_positions={}):

    pass


def convert_shape_format(shape):

    pass


def valid_space(shape, grid):

    pass


def check_lost(positions):

    pass


def get_shape():

    pass
```

```python
def draw_text_middle(text, size, color, surface):

    pass


def draw_grid(surface, row, col):

    pass


def clear_rows(grid, locked):

    pass


def draw_next_shape(shape, surface):

    pass


def draw_window(surface):

    pass


def main():

    pass
```

```python
def main_menu():

    pass



main_menu()  # start game
```

Part #2: Game Design

I have divided the code into various sections for easier understanding. You need to keep all this code in one script file.

Class for Piece Shapes

We are going to create different shapes and it's better to create a class for all shapes that defines all the common properties and functions.

```python
class Piece(object):

    rows = 20  # y

    columns = 10  # x



    def __init__(self, column, row, shape):

        self.x = column

        self.y = row
```

```python
        self.shape = shape

        self.color = shape_colors[shapes.index(shape)]

        self.rotation = 0  # number from 0-3
```

Create a Game Grid

This piece of code creates the visible gaming area. The randomly generated shapes will occupy this grid according to the user input.

```python
def create_grid(locked_positions={}):

    grid = [[(0,0,0) for x in range(10)] for x in range(20)]


    for i in range(len(grid)):

        for j in range(len(grid[i])):

            if (j,i) in locked_positions:

                c = locked_positions[(j,i)]

                grid[i][j] = c

    return grid
```

Randomizing Shape Generation

This code generates different shapes randomly. Note that we already set all the shapes that will be generated during initialization.

```
def get_shape():

    global shapes, shape_colors


    return Piece(5, 0, random.choice(shapes))
```

Game Grid Build

This part of script builds the grid where the game will be played.

```
surface.fill((0,0,0))

    # Tetris Title

    font = pygame.font.SysFont(\'comicsans\', 60)

    label = font.render(\'TETRIS\', 1, (255,255,255))


    surface.blit(label, (top_left_x + play_width / 2 - (label.get_width() / 2), 30))


    for i in range(len(grid)):
```

```
        for j in range(len(grid[i])):

            pygame.draw.rect(surface, grid[i][j], (top_left_x + j* 30,
top_left_y + i * 30, 30, 30), 0)


    # draw grid and border

    draw_grid(surface, 20, 10)

    pygame.draw.rect(surface,   (255,   0,   0),   (top_left_x,
top_left_y, play_width, play_height), 5)

    pygame.display.update()
```

Create Game Loop

This is the main loop that will be constantly running looking for events and invoking necessary actions. The "while" loop helps create an infinite iteration that will only break if we lose or win the game.

```
def main():

    global grid


    locked_positions = {}  # (x,y):(255,0,0)

    grid = create_grid(locked_positions)
```

```python
    change_piece = False

    run = True

    current_piece = get_shape()

    next_piece = get_shape()

    clock = pygame.time.Clock()

    fall_time = 0


    while run:

        for event in pygame.event.get():

            if event.type == pygame.QUIT:

                run = False

                pygame.display.quit()

                quit()


            if event.type == pygame.KEYDOWN:

                if event.key == pygame.K_LEFT:

                    current_piece.x -= 1
```

```python
            if not valid_space(current_piece, grid):

                current_piece.x += 1


        elif event.key == pygame.K_RIGHT:

            current_piece.x += 1

            if not valid_space(current_piece, grid):

                current_piece.x -= 1

        elif event.key == pygame.K_UP:

            # rotate shape

            current_piece.rotation = current_piece.rotation + 1 % len(current_piece.shape)

            if not valid_space(current_piece, grid):

                current_piece.rotation = current_piece.rotation - 1 % len(current_piece.shape)


        if event.key == pygame.K_DOWN:

            # move shape down

            current_piece.y += 1
```

```
        if not valid_space(current_piece, grid):

            current_piece.y -= 1



    draw_window(win)
```

Part #3: Building the Game

Add Visuals to the Game Grid

We are going to add grid lines so players will know how much space a shape takes and how much the shape can be moved.

```
def draw_grid(surface, row, col):

# This function draws the grey grid lines that we see

    sx = top_left_x

    sy = top_left_y

    for i in range(row):

        pygame.draw.line(surface, (128,128,128), (sx, sy+ i*30), (sx
+ play_width, sy + i * 30))  # horizontal lines

        for j in range(col):

            pygame.draw.line(surface, (128,128,128), (sx + j * 30, sy),
(sx + j * 30, sy + play_height))  # vertical lines
```

Track Position of Shape in the Grid

During the game, we must be able to track the position of the shapes. Here's the code to do that.

```python
def convert_shape_format(shape):

    positions = []

    format = shape.shape[shape.rotation % len(shape.shape)]


    for i, line in enumerate(format):

        row = list(line)

        for j, column in enumerate(row):

            if column == \'0\':

                positions.append((shape.x + j, shape.y + i))


    for i, pos in enumerate(positions):

        positions[i] = (pos[0] - 2, pos[1] - 4)


    return positions
```

Check if Certain Grid Space is Valid

To make sure the shape doesn't move into a space that's already taken or is out of the available grid, we have to write a function. Here's the code to check grid space validity.

```
def valid_space(shape, grid):

    accepted_positions = [[(j, i) for j in range(10) if grid[i][j] ==
(0,0,0)] for i in range(20)]

    accepted_positions = [j for sub in accepted_positions for j in
sub]

    formatted = convert_shape_format(shape)


    for pos in formatted:

        if pos not in accepted_positions:

            if pos[1] > -1:

                return False


    return True
```

Check if Game is Lost

We have to constantly check if user has lost the game. Here's the function code to check that.

```python
def check_lost(positions):

    for pos in positions:

        x, y = pos

        if y < 1:

            return True

    return False
```

Modify Game Loop

Remember we created the main game loop in Part #2? We need to add more codes to it. Add the following code block right after the "while run:" line.

```python
fall_speed = 0.27


    grid = create_grid(locked_positions)

    fall_time += clock.get_rawtime()

    clock.tick()


    # PIECE FALLING CODE

    if fall_time/1000 >= fall_speed:
```

```
        fall_time = 0

        current_piece.y += 1

        if    not    (valid_space(current_piece,    grid))    and
current_piece.y > 0:

            current_piece.y -= 1

            change_piece = True
```

Before the "draw_window(win, grid)" line, add the following lines of code.

```
shape_pos = convert_shape_format(current_piece)


    # add color of piece to the grid for drawing

    for i in range(len(shape_pos)):

      x, y = shape_pos[i]

      if y > -1: # If we are not above the screen

        grid[y][x] = current_piece.color

    # IF PIECE HIT GROUND

    if change_piece:

      for pos in shape_pos:
```

```
    p = (pos[0], pos[1])

    locked_positions[p] = current_piece.color

    current_piece = next_piece

    next_piece = get_shape()

    change_piece = False
```

After the "draw_window(win, grid)" line, add the following lines of code.

```
# Check if user lost

    if check_lost(locked_positions):

        run = False
```

Part #4: Adding Features to The Game

Show Next Shape on the Side

We want to show the next shape that will enter the grid from the top on the right side of the grid. Here's a function to do that.

```
def draw_next_shape(shape, surface):

    font = pygame.font.SysFont('comicsans', 30)

    label = font.render('Next Shape', 1, (255,255,255))
```

```python
    sx = top_left_x + play_width + 50

    sy = top_left_y + play_height/2 - 100

    format = shape.shape[shape.rotation % len(shape.shape)]


    for i, line in enumerate(format):

        row = list(line)

        for j, column in enumerate(row):

            if column == 'o':

                pygame.draw.rect(surface, shape.color, (sx + j*30, sy +
i*30, 30, 30), 0)



    surface.blit(label, (sx + 10, sy- 30))
```

We need to call this function in the main() so it will constantly
run.

```python
# This should go inside the while loop right BELOW
draw_window()

# Near the end of the loop

draw_next_shape(next_piece, win)
```

pygame.display.update()

Clear Filled Row

When a grid row is completely filled, it must be removed from the grid and all other rows shifted down. Here's the function code to do that.

```python
def clear_rows(grid, locked):

    # need to see if row is clear then shift every other row above
down one



    inc = 0

    for i in range(len(grid)-1,-1,-1):

        row = grid[i]

        if (0, 0, 0) not in row:

            inc += 1

            # add positions to remove from locked

            ind = i

            for j in range(len(row)):

                try:
```

```
            del locked[(j, i)]

        except:

            continue

    if inc > 0:

        for key in sorted(list(locked), key=lambda x: x[1])[::-1]:

            x, y = key

            if y < ind:

                newKey = (x, y + inc)

                locked[newKey] = locked.pop(key)
```

We also need to call this function from the main() function. Add the following lines of code at the end of "if change_piece:" block.

call four times to check for multiple clear rows

```
        clear_rows(grid, locked_positions) # < ---------- GOES
HERE
```

Part #5: Raise Difficulty

In a Tetris game, we can increase the game difficulty by increasing the speed with which the shapes fall down the grid.

In the main() function, just before the "while run:" line, add the following line of code.

level_time = 0

Now, add the following line of code at the start of "while run:" loop.

level_time += clock.get_rawtime()

```
    if level_time/1000 > 5:

        level_time = 0

        if level_time > 0.12:

            level_time -= 0.005
```

Part #6: Game Build

The following lines of code will go at the end of the script.

win   =   pygame.display.set_mode((s_width,   s_height)) pygame.display.set_caption(\'Tetris\')

main()

We can add many more features to the game to offer a better gaming experience. Here are some ideas:

1. Add a scoring system. For example, award 10 points for each filled row cleared from the grid
2. Add a main menu to the game
3. Add an exit screen with score etc. summary after the game stops running

The above example is just the beginning when it comes to creating games with Python. Well-known games including Battlefield 2 and The Sims 4 utilize the power of Python to create a highly immersive gaming experience. If you want to become a game developer and don't have the capital to invest in proprietary platforms, Python can help you become famous!

Creating an Executable Distribution of Software

We have to install an external library to create distributable ".exe" package for any Python script.

pip install pyinstaller

Using the command prompt, navigate to the directory where the Python script is present. Let's say our script file name is "appComp.py". We have to run the following command.

pyinstaller --onefile appComp.py

The process can take a few minutes to complete depending upon your script. It automatically takes care of any libraries you added in the script. Once it's finished, a few new folders and files will be present in the folder where your Python script was present. Open the "dist" folder and you will see the "appComp.exe" file. Easy!

Conclusion

Python has created a revolution in the programming world due to its open source nature. The rise of data analysis and machine learning have also made Python very relevant. It offers easier syntax but doesn't compromise in power or speed. All these qualities have made the language ideal for people who are not programmers but want to create scripts to achieve specific objectives.

Python 2.7 is the most popular version even though the support is going to end very soon. Code written in Python 2.7 is not forward compatible with Python 3.0. It definitely created a dilemma for companies who have relied on Python for years. It's not easy to port millions of lines of code. Due to this reason, Python 2.7 End of Life (EOL) has been extended several times. But, it's certain it will not be extended this time.

There are various books and YouTube channels that you can refer to if you want to build upon the skills learned in this book. Python is a great programming language if you want to freelance or start your own programming firm. I hope your Python journey is fun and full of rewards!

References

Eric Matthes, Python Crash Course, 2nd edition. San Francisco, CA: No Starch Press Inc., 2019.

James R. Parker, Python: An Introduction to Programming. Dulles, VA: Mercury Learning and Information, 2017.

Ryan Turner, Python Programming: 3 Books in 1: Ultimate Beginner's, Intermediate & Advanced Guide to Learn Python Step-by-Step. James C Anderson, 2018.

Steven Samelson, Python Programming: A Step-by-Step Guide From Absolute Beginners to Complete Guide for Intermediates and Advanced. Steven Samelson, 2019.

www.ingramcontent.com/pod-product-compliance
Lightning Source LLC
Chambersburg PA
CBHW071114050326
40690CB00008B/1222